AI Revolution

Mastering AI for Personal and Organizational Growth

Published 2024 by River Publishers
River Publishers
Alsbjergvej 10, 9260 Gistrup, Denmark
www.riverpublishers.com

Distributed exclusively by Routledge
605 Third Avenue, New York, NY 10017, USA
4 Park Square, Milton Park, Abingdon, Oxon OX14 4RN

AI Revolution: Mastering AI for Personal and Organizational Growth / Tero Ojanperä.

Cover: Brent Beckley with AI

Routledge is an imprint of the Taylor & Francis Group, an informa business

ISBN 978-87-7004-232-1 (hardback)
ISBN 978-87-7004-648-0 (paperback)
ISBN 978-87-7004-640-4 (online)
ISBN 978-8-770-04631-2 (ebook master)

AI Revolution

Mastering AI for Personal and Organizational Growth

Tero Ojanperä

River Publishers

Routledge
Taylor & Francis Group
NEW YORK AND LONDON

Contents

Foreword to the English Edition

Dear Readers, welcome to the English edition of "AI Revolution – The Handbook." As we stand at the threshold of a major technological shift, this book serves as your guide through the diverse world of artificial intelligence and its wide-ranging applications.

Agent Technology

The original writing of this book in Finnish took about two months, greatly assisted by AI technology. However, the translation process was astonishingly quicker, thanks to the advancements in AI. By simply inputting the Finnish manuscript into a large language model, the English version was generated in just a few seconds. With just a few factual corrections from my side, the translation was completed. Of course, you will notice that it's not perfect, which highlights the need to continue developing large language models and the importance of human effort. We also prompted the large language model to produce the layout. Interestingly, the time-consuming part was adapting the figures and AI chat examples, which required more meticulous and manual effort than the textual translation. I also created an audio book, read by AI but with my own voice. The world of publishing is definitely going to change radically with AI.

The path to translating this work was an enlightening one. Jani Pitkänen and Heikki Lehmuskoski from Boost with AI played a crucial role, skillfully handling the translation and ePub layout design to create an English version that speaks clearly and effectively. Thank you also, Reima Kokko, for help with the graphics.

The journey of AI transformation is just beginning, and it will take time for companies to understand that it's not about adding AI on top of existing products and services, but about reimagining workflows, products, and processes. However, the effort will undoubtedly be worth it. The advancements in AI have been rapid in the months since the Finnish version of my book was released. Notable developments include the launch of Google's Gemini model, Pika's text-to-video model, the integration of Dalle-3 into ChatGPT, Microsoft rolling out its AI copilots for Office tools, and the development of the Finnish "Poro" (Reindeer) open-source model by Silo AI and the University of Turku, among many others.

The pace is so fast that it's hard to keep up with all the developments. However, we should not be so worried about the short term AI news but focus on the long term. This is a revolution that will last for the next decade, thus focus on the development of your own AI skills, your organization's AI strategy and transformation as well as how AI will impact society and how we can influence it.

Äkäslompolo, Lapland

9 January 2023

Foreword

On 7 February 2017, I stepped out simultaneously with Juha Hulkko and Peter Sarlin after the Prime Minister's artificial intelligence seminar. Juha, a successful technology entrepreneur, had just introduced me to Peter, a young artificial intelligence researcher and entrepreneur. There is something big and exciting in the moment. We have all presented our views on the state of artificial intelligence and what Finland should do at the seminar. The idea arises that we should also do something concrete.

Our journey with artificial intelligence had already begun earlier. Under Juha's leadership, we had gathered to discuss why artificial intelligence is not talked about enough in the context of the "Intelligence is Action" project. This discussion led to a seminar on artificial intelligence convened by Prime Minister Juha Sipilä. After the seminar, things started to happen. Minister Mika Lintilä set up a steering group to prepare a proposal for Finland's AI program, and with Business Finland's funding aimed at AI companies, there was a significant effort to accelerate AI development.

Peter and I decided to start developing artificial intelligence projects that we could use to approach our customers. Over the coming months, the decision to establish Silo AI was made. The founding team included, in addition to Peter, Juha and myself, startup entrepreneur Ville Hulkko, Johan Kronberg who had a long career at PwC, and Associate Professor Kaj-Mikael Björk. The opportunity was too good to pass up. Artificial intelligence would become a force for change, just like the internet and mobile had been before.

Since then, Silo AI has grown into a 300-person artificial intelligence company that serves as an AI partner in customer product development. The most recent expansion took place through the acquisition of the artificial intelligence company Combient Mix, founded by the Wallenbergs. From the beginning, Silo's goal has been to build the best workplace for AI experts. Half of our staff are indeed PhDs – we strongly invest in continuous learning and culture development.

Silo AI's vision has been to create artificial intelligence for people (AI for people). The values we created in the summer of 2018 (build bonds, keep learning, ask why, be good) have been our concrete guidelines as the company has grown. One must build bonds with other people, constantly learn, ask "why", and act for the good of others. As the AI revolution accelerates, these values are more relevant than ever.

My entire career, since I started my diploma thesis at Nokia in 1990, has been deeply linked to technology. My observation is that disruptions actually happen slowly. The latest revolution in artificial intelligence did not start with ChatGPT, nor did the history of smartphone technology start with the iPhone. They were turning points that catalyzed the industry in a new direction, but groundwork had been done for a long time. The industry change from a turning point usually takes years. Businesses and society have time to adapt, but often it is not done because the signs of the times are not read correctly.

That's exactly why I wrote this book. Five years after the founding of Silo AI, ChatGPT catalyzed the second wave of artificial intelligence that will last for the next decades. Generative AI significantly affects individuals, businesses, and society. It shapes our language and culture. For this reason, Silo AI also founded a new company focused on generative artificial intelligence, SiloGen.

A lot has been written about artificial intelligence, but I wanted to address the topic from various perspectives. I want to help the reader navigate through the whirlwinds of this revolution and make better decisions. Whether you are a corporate leader, an end user, or a public sector decision maker, artificial intelligence will change your organization and your work, and its impacts cannot be ignored.

Finland and Europe need to act now, build momentum and invest in the future. That, in turn, requires that we understand what is at stake. Creating this understanding is the goal of my book.

The artificial intelligence sector is developing at a rapid pace, with news of innovations coming almost daily. In the future, I will be writing about the development of artificial intelligence on the website tekoalynvallankumous.fi.

Thank you to the Alma Talent team (Maria Ampiala, Riikka Rajapuro, and Sari Hawkins) for embarking on this intensive project. Publishing editor Hanna Virusmäki did excellent work, and together with artificial intelligence, we formed a seamless team. Valuable feedback on the manuscript was provided by Pia Relander-Paavola, Laura Niemi, Juha Hulkko, Timo Lappi, and Pekka Koponen. Thanks to the Silo AI team and customers, from whom I have learned a lot, as well as my colleagues at Aalto, FCAI, and other universities. Thanks also to the Facebook AI group for inspiration and to friends and colleagues who have supported me during the writing process.

I dedicate this book especially to my beloved wife Tiina and our three sons Eerik, Elias, and Eemil. Yet another summer vacation, numerous weekends, car trips, even Formula 1 race breaks were spent writing this book. Thank you for your support and understanding.

At Silo AI, it is often asked which of the company's values are everyone's favorite. My favorite value is "keep learning" – continuous learning. I hope you enjoy reading and learn something new.

In Espoo,

30 August 2023

About the Author

Dr. Tero Ojanperä is a visionary technologist and pioneering entrepreneur. As the Chairman and Co-Founder of Silo AI, the largest private AI lab in the Nordics, he is a driving force behind the AI revolution, shaping the future of technology and business. In addition to his groundbreaking work at Silo AI, Tero holds the position of Professor of Practice in Intelligent Platforms at Aalto University, Finland. Previously, he was Nokia's CTO and Chief Strategy Officer and served on its executive board.

He is a co-author of the influential book "Platform Strategy: Transform Your Business with AI, Platforms and Human Intelligence" which provides a blueprint for businesses navigating the digital age. His accomplishments have been recognized globally, with a nomination as a Young Global Leader by the World Economic Forum in 2006 and a ranking as the 7th Most Creative Person in Business in 2009 by Fast Company.

He enjoys Olympic weightlifting, wooden boats, triathlon, and skiing in his spare time.

INTRODUCTION

The Revolution Spreading at the Speed of Light

The New York Times reported in November 2022 that OpenAI's engineers were somewhat concerned about what they should do next. Should they wait for the completion of the new extensive language model GPT-4, or launch the ChatGPT tool with the unfinished GPT-3 model. Waiting might lead to competitors reaching the market first, but an unfinished model could cause a PR disaster if it didn't work as desired.

The fear turned into energy when the decision to release GPT-3 was made. ChatGPT was launched on 30 November, and by 4 December, it broke the one million user milestone. And even better was to come. ChatGPT reached one hundred million users just two months after its release, faster than any other app before. With that, we moved into the era of generative artificial intelligence. Now, massive AI models generate text, images, music, moving images, and even 3D models in a way that was not previously believed to be possible for machines.

Although ChatGPT became the flag bearer of the revolution, the AI upheaval had already begun earlier. In 2021, the Runway company and researchers at the University of Munich completed an artificial intelligence model called Stable Diffusion, which created images from text commands. British startup Stability AI, on the other hand, paid for the computational costs of training the model. In August 2022, Stability AI commercially launched the Stable Diffusion model, transforming it from an academic research project into a global phenomenon. "Wow, AI is pretty cool!" would be pretty close to what people experienced at that time.

Generative artificial intelligence enabled a conversational interface. Interaction with artificial intelligence occurs in natural language; these applications are spoken or written to like humans. Machines and humans communicate seamlessly, without friction. This is the “iPhone moment” of artificial intelligence. The conversational interface democratizes artificial intelligence, which is why it is so revolutionary.

Generative artificial intelligence applications are coming to the market at an increasingly rapid pace. The number and diversity of applications are driven by the continuous development of AI models, their ease of implementation, and the race between startups and tech giants for dominance in new technology.

The revolution begins when people are ready to break old systems and create new ones. The revolution shakes up power structures, political systems, and social norms thoroughly and quickly. This is exactly what is happening as artificial intelligence spreads at the speed of light to everyday applications. It challenges education, teaching and work models, and changes societal norms regarding what is allowed and what is not. Pioneers do not wait, they do not consider whether something is allowed or not, they start acting.

Frontlines of the Revolution

The battle for the dominance of the new artificial intelligence is being fought on many fronts. In the constant stream of AI news, it can be difficult to grasp what is happening. That’s why it’s good to distinguish the development of AI technology from AI applications.

ChatGPT, for example, is an application built on top of an artificial intelligence model. ChatGPT currently uses the GPT-3.5 model in its basic version and the more capable GPT-4 model in its paid version. Meta, Facebook’s parent company, has brought to market an open-source model called LLaMA, on top of which applications can be built. These models are referred to as base models. They are constantly evolving, creating new opportunities for different application developers and users.

Text-to-image models are constantly evolving: AI models that create videos from text have already been introduced to the market. And in addition to text, the models are starting to understand images as well; this is referred to as multimodality. Companies are developing their own specialized models, such as TietoEvry for healthcare.

If one company controlled artificial intelligence models, all applications would be dependent on it. Fortunately, innovations are happening. I will go through artificial intelligence models, technology, and their development in more detail in Chapter 1.

Artificial intelligence models are often used through an interface (API, application programming interface), which is why application developers do not need to know everything about artificial intelligence technology, as the necessary capabilities can be incorporated through interfaces. That's why we see rapid development in new applications utilizing artificial intelligence.

Companies are building entirely new applications on top of generative AI models and, on the other hand, they are also renewing old applications by introducing generative AI into them. Examples are Microsoft's 365 Copilot or the AI assistant of the language training application Duolingo.

From Hype to Bright Rise

The hype around artificial intelligence has led many to question whether it's a bubble. This can be examined with Gartner's hype cycle, which is a popular way to understand the development of new technology. It consists of five stages: technology launch, inflated expectations, trough of disillusionment, slope of enlightenment, and plateau of productivity. The hype cycle is also suitable for examining artificial intelligence.

The release of ChatGPT and other AI applications has raised hype and excitement, as a result of which expectations have swelled. A lot is expected from the technology. AI is imagined to solve all problems, but its real capabilities may not (yet) meet expectations. For this reason, this phase is usually followed by disappointment. When technology gradually begins to fulfill its promises, a new rise follows, and productivity improves. At the stage of productivity leveling off, the technology is fully accepted and widely used. It has proven its value and fulfilled its promises.

After the initial excitement, it has been noticed that the new artificial intelligence technology also poses many challenges. Artificial intelligence hallucinates or invents stories "from its own head", meaning it doesn't give truthful answers. This can prevent the use of AI in demanding applications. Specialized models also need diverse data, the acquisition of which can be slow and expensive. AI was expected to revolutionize internet search, but it doesn't work like a traditional search engine. In the summer of 2023, it was reported that the quality of ChatGPT had deteriorated and user numbers had started to decline.

All of these are natural parts of the development of new technology. The tech bubble of the 2000s led to disappointments, but it was followed by a rise that gave us Google, Facebook (Meta), Amazon, and other tech giants. Technological disruption takes time, but eventually, new winners stand out from the rest. The AI revolution also takes time, and disappointments and failures are a natural part of it. But that's why the transition requires a long-term approach from companies and society.

OpenAI – Background and Strategy

Elon Musk and LinkedIn founder Reid Hoffman were involved in founding OpenAI in 2015. OpenAI was initially a non-profit company aiming to advance the development of artificial intelligence. However, Musk ended up in disputes with other founders and left the company three years later. In 2019, OpenAI transitioned from a non-profit organization to a capped for-profit organization, and one of the company's founders, Sam Altman, rose to the company's leadership.

Interestingly, Altman did not want ownership of OpenAI. He explained his decision by saying that this way he could be an impartial AI developer. He was also already super rich, as he was also the founder of the legendary business development company Y Combinator. As OpenAI began to pursue profit, its business model changed. Despite its name, there is nothing open about the company: OpenAI does not disclose the operation of its AI models or its training data for others to see.

In addition to ChatGPT, OpenAI has several artificial intelligence models and products: DALL·E generates images from text, Codex is an AI model specialized in code generation, and Whisper is a speech recognition model. The company is constantly striving to develop new artificial intelligence technology and expand its offerings. OpenAI's programming interface (API) provides applications with an easy way to use its AI models and build new applications. Applications can be connected to ChatGPT through plugin functionality. The overall goal is to expand OpenAI's ecosystem; as in technology, the player who grows the fastest and gets the most users usually wins.

OpenAI has also sought to promote the spread of its technology through a startup fund and by creating partnerships with different user groups. A partnership has been formed, for example, with the American Journalism Project (AJP) that focuses on renewing local news with the help of artificial intelligence.

OpenAI's strategy is to introduce its technology to new user groups. In August 2023, it launched ChatGPT Enterprise for businesses, which offers features that enhance data security and privacy.

Microsoft is a partner and significant owner of OpenAI. In 2019, it invested a billion dollars in the company, and in 2023, there were reports of an additional investment of ten billion dollars. Microsoft's goal is to take advantage of OpenAI's technology at the forefront.

Tech Giants Apply Artificial Intelligence

Technology analyst Benedict Evans ponders in his newsletter in March 2023 (which is worth subscribing to: www.ben-evans.com), how artificial intelligence spreads as established and new companies adopt it. Established companies add generative AI features to their existing platforms. These companies and their applications have a stable user base, so artificial intelligence spreads quickly. The AI-enhanced Photoshop is a good example.

The world's largest companies have dramatically brought artificial intelligence into their products and services. Google, for example, paints a future of an AI-empowered Gmail, which would not only organize emails but also help draft, respond, summarize, and prioritize messages. With the Google Docs application, the writer would no longer need to stare at a blank page wondering where to start. They would just input the topic they want to write about, and the AI would create a preliminary draft. Google Slides, on the other hand, would generate fully prepared slides for the user. Google has also created a competitor for ChatGPT, Gemini.

If the user is a captain, Microsoft provides him with a co-pilot, Copilot, to assist with Office tools. It is an AI partner that supports work. Behind everything are OpenAI's AI models, such as GPT-4 and DALL·E. With their help, the user gets assistance in writing in Word and creating images in PowerPoint. The brand-new Business Chat can even be asked to update the company's product strategy – the program creates a situation update based on morning meetings, emails, and chat threads.

The basic principle of Word is the same as 40 years ago when version 1.0 hit the market in 1983. However, each technological change also offers new players the opportunity to enter the market. Startups often approach new technology in a different way than established players and create tools that are built from the ground up to take full advantage of the new technology's potential – instead of adding it to existing tools.

According to Evans, the challenge for established companies is to balance existing products and new innovations, such as artificial intelligence, in their offerings to stay ahead of competitors. Balance is a delicate state – history shows that those who manage to navigate the transition effectively can achieve long-term success.

Microsoft Office has been declared dead many times, but it has over 300 million users worldwide. If Microsoft succeeds in elevating artificial intelligence as part of the Office user experience, it will remain relevant as a software suite and share the benefits of artificial intelligence with a very large group at once. As we move further into the AI era, we will see an exciting mix of evolution and revolution.

Generative Fill is a new feature in Adobe Photoshop that utilizes Adobe's Firefly AI model to enhance design and creation processes. It provides users with the ability to add, expand, or remove content from images in seconds with simple text comments. Generative Fill has been trained on hundreds of millions of professional, licensed, and high-resolution images from Adobe's Stock image service: the program does not create content based on the intellectual property rights of other people or brands.

In addition, Generative Fill enables the creation of new content in generative layers, which accelerates the experimentation of creative possibilities and the reversal of effects if necessary without compromising the original image. It also provides the opportunity to produce concepts and create high-quality content even faster.

Meta has profiled itself into the Metaverse in recent years. Although the "metaverse" seems old-fashioned during the hype of generative AI, Meta has been doing artificial intelligence research for a long time. It launched an open-source model, Llama 2, in the summer of 2023, which other players can also use commercially. Meta's goal is likely to prevent another tech giant from gaining too strong a position in the field of artificial intelligence. Services similar to ChatGPT, such as Chat with Llama 2[1] and Perplexity-chat[2] have already been built on the Llama 2 model.

To prevent the game from appearing too simple, Microsoft and Meta announced a strategic partnership regarding Llama 2. This may seem strange, given that Microsoft has invested billions in OpenAI. But Microsoft's main strategy is to increase the use of its Azure cloud service, and in that sense, it doesn't matter to the company whether GPT-4 or Llama 2 is running there. Microsoft just needs to ensure that it has access to the latest technology.

---

[1] Chat with Llama 2: https://www.llama2.ai/
[2] Llama Chat, Perplexity Labs: llama.perplexity.ai

When following the announcements of tech giants, one must remember that they are just that, announcements. Companies strive to position themselves in the minds of users, businesses, and investors as pioneers of technology by releasing new features in advance, this is also the case with artificial intelligence. Over time, we will see how the application of artificial intelligence moves from hype through disappointments to a new rise.

Startup Investments on the Rise

Although startup investments significantly decreased in 2023, investments in generative artificial intelligence have been on the rise. As the hype grows, investors rush to shop. Everyone wants to be part of the hottest trend.

OpenAI has made several very large investment rounds, they are in the hundreds of millions of dollars. Anthropic received funding of about 300 million dollars in early 2023. The company is developing Claude AI assistant and models. Claude can be used like ChatGPT, or its models can be integrated into other applications.

Adept received a $350 million investment to develop generative artificial intelligence for various digital tools. The user can write what task they want to do, and the model performs the task by translating the natural language description into commands that different applications understand.

CB Insights monitors the generative AI market. According to them, currently over three hundred startups are developing AI technology, solutions, and applications, and more are coming all the time.

Artificial Intelligence Creates New Ways to Use Applications

Generative artificial intelligence is used in more and more applications. One could say that almost all applications will benefit from the capabilities of the new AI. Particularly interesting, however, are applications that create entirely new uses. However, these are difficult to imagine in advance. For example, we did not predict the rise of search engines or social media. In 2000, there was no Twitter, Facebook or Netflix. We also didn't imagine that we would be constantly glued to our phones.

A year ago, I wouldn't have imagined having a conversation with a PDF document, but now it's possible with ChatGPT using the AskPDF or ChatwithPDF plugins. The virtual assistant Be My Eyes helps visually impaired people interact with their environment. In it, the user can ask anything about the pictures taken by the program and thus get help with various everyday tasks.

The Duolingo language application offers a personal way to learn languages. With the help of artificial intelligence, the student can ask the program to explain the answer they received. Artificial intelligence can also take on the role of a waiter, for example, when the user is learning to operate in a restaurant with a new language.

Artificial intelligence helps companies manage information in new ways. It is also applied to the development of new drugs. The possibilities seem limitless. However, it is difficult to predict where the real breakthroughs will occur. The only certainty is that the revolution has begun and it is changing the way we operate. Just as cell phones freed us from location and touch screen and smartphone democratized the internet, generative artificial intelligence creates pervasive intelligence everywhere.

The Multifaceted Impacts of the AI Revolution

This book is a practical guide to the AI revolution. It helps the reader understand the underlying technology, its use, and its impact on professions and productivity. The book also discusses legal issues raised by generative AI, societal impacts, and future vision.

Chapter 1 is a profound overview of the technology behind generative artificial intelligence, namely the basics and operation of large AI models. After reading it, you will understand why such models can produce intelligent content and solve complex problems. You will also understand why they don't always work the way we want and what challenges are associated with the technology.

Chapter 2 moves on to practice: you will learn how to multiply your productivity by instructing or prompting artificial intelligence in different ways. The chapter particularly goes through the instruction of ChatGPT but also the implementation of Google Bard and similar other tools, such as Claude and Llama 2. The use of programs that produce images from text, such as Midjourney and DALL·E, is also delved into.

Chapter 3 opens up how artificial intelligence affects the everyday life of a white-collar worker, changing, destroying, and creating jobs. Whether you are an expert, manager, coder, student, teacher, doctor, lawyer, or creative professional, you cannot escape the effects of artificial intelligence. In this chapter, I explain what the AI revolution requires from individuals and companies, and what strategic capabilities the change requires.

Chapter 4 describes how artificial intelligence increases productivity and why the leap in productivity is greater than that generated by the internet. I also describe what the role of society, workers, and companies is in implementing the productivity leap.

Chapter 5 discusses how generative artificial intelligence is fundamentally reshaping society. It provides new tools for producing, analyzing, and sharing knowledge and will lead to new innovations that are difficult to predict. From a societal perspective, artificial intelligence is becoming a new civic skill.

In Chapter 6, I discuss how extensive language models integrate into the physical world and how this affects the development of intelligent robots. Robots are becoming learners, and they will soon be able to perform tasks they have not previously encountered. I also explain how intelligent robots are trained using videos.

Chapter 7 deals with legal issues raised by generative artificial intelligence, such as copyrights and AI regulation. The chapter discusses whether the output of AI is a work, who owns its rights, and what material can be used to teach AI. No one currently knows the future of AI regulation and future jurisprudence, but four scenarios can be created to help understand how to navigate the jungle of legal issues.

Chapter 8 considers how generative artificial intelligence affects language, culture, and even the worldview. The way we use language reflects our identity, values, and attitudes. For this reason, large language models not only learn to understand language, but they also adopt various societal and cultural prejudices. How do we ensure that we control the change from this perspective? The ideologies of AI developers affect how the outputs of AI shape our thoughts. In the hands of a dictator, AI can be a very dangerous weapon. The chapter delves into how we preserve our culture and values in the AI revolution.

Chapter 9 directs thoughts to the future – can superintelligence, for example, be born?

Artificial intelligence singularity refers to a hypothetical future point when artificial intelligence systems reach a level of intelligence and capabilities that surpass human possibilities and understanding. Could we end up in this situation? How would it affect our world? Directing artificial intelligence is a process in which the goals and actions of artificial intelligence are aligned with human values, interests, and goals. The chapter discusses how this is done and how safe artificial intelligence can be developed.

The AI revolution sparked by ChatGPT is just the beginning: this work is a handbook for you who want to keep up with the rapid pace of AI development.

CHAPTER

1

AI Primer: How a Superhero is Born

I'm finalizing this chapter in August 2023. It feels difficult because so much is happening on the artificial intelligence front. Just when I think I'm done, I notice a new publication about some new and even better artificial intelligence model. Frustrating, but at the same time exciting!

Artificial intelligence services, such as ChatGPT, are the superheroes of the new era. We were amazed when ChatGPT was able to answer questions, summarize an article in milliseconds, code, write poems, song lyrics, and much more.

ChatGPT solves tasks that require in-depth knowledge, analysis, and creativity, and it performs them skillfully. It can figure out the principles of quantum physics and create an epic poem. But if the instructions given to it are unclear, its logic gets completely mixed up. Sometimes it may also invent things, mixing facts and fiction so smoothly that it starts to "hallucinate" new worlds, events, and characters. It feels like a superhero has eaten mushrooms.

The number of superheroes is also growing all the time: Google Gemini, Midjourney, DALL·E, Stable Diffusion, Claude, and others have entered the market. To understand that artificial intelligence is not a conscious being or "someone" who develops on their own and has superhuman abilities, we also need to understand technology. Behind these superheroes are not supernatural powers, but gigantic basic models of artificial intelligence that have been trained with a massive amount of data.

In this chapter, I talk about how generative artificial intelligence was born, how modern AI models work, and how they are tuned for different purposes. This chapter may feel heavy for a non-technical reader, but you can read it alongside Chapter 2.

1.1 A Brief History of Artificial Intelligence and Basic Concepts

Artificial intelligence as a field of science was born when computer science researchers began in the 1940s to investigate whether machines could be taught to think like humans.

In the summer of 1956, researchers Marvin Minsky, John McCarthy, Claude Shannon, and Nathaniel Rochester organized a multi-day seminar for field researchers with a grandiose name: The Dartmouth Summer Research Project on Artificial Intelligence. John McCarthy coined the term "artificial intelligence" and persuaded others to accept its use. At this seminar, researchers Allen Newell, Herbert Simon, and Cliff Shaw introduced their developed Logic Theorist computer program, which is considered the world's first artificial intelligence program, as it was capable of automatic reasoning. The first artificial intelligence programs were simple and mainly based on programmed rules.

The years following the Dartmouth seminar were a time of enthusiasm. Computers were able to solve complex mathematical problems that were not believed to be possible for machines. Researchers were optimistic and predicted that a fully intelligent machine would be developed in less than 20 years. This optimism led to significant investments.

When discussing the development of artificial intelligence, the term "spring" is often used to describe experiences of success; "winter" refers to disappointments that the field has encountered. After the "spring" of the late 1950s, artificial intelligence experienced its first "winter" in the early 60s, when it was realized that machines operating on the basis of programming rules were not able to solve more complex problems. This led to the birth of a new approach, machine learning.

Machine learning is a method where artificial intelligence learns from data. It is divided into three main types: supervised learning, unsupervised learning, and reinforcement learning. The machine learning of artificial intelligence requires a lot of repetition, much like the learning process of a child or a puppy – in its learning process, artificial intelligence is constantly facing something new.

In supervised learning, the machine is taught to recognize elements from, for example, images or texts. Think of CAPTCHA codes or image verification, which aim to ensure that the user of the site is a human. Traffic lights or cars are recognized from images; each such time helps artificial intelligence to recognize traffic signs better in the future. In this way, each of us is just one cog in the teaching machinery of artificial intelligence.

In unsupervised learning, artificial intelligence learns from data without predefined classes or models. The machine learns on its own and by examining the data. So no one tells it that there are traffic lights or cars in the pictures. At its best, unsupervised learning can, for example, help find new features or genetic groups in human genomes that are essential in diagnosing or treating diseases.

In reinforcement learning, the machine learns by trying and learning from its mistakes. One could think that a self-driving car is like a puppy learning to walk on a leash. The car, otherwise fully ready for travel, does not yet understand traffic rules, just as the puppy does not know how to walk on a leash. In reinforcement learning, the puppy gets treats when it makes the right choices – similarly, the self-driving car gets positive feedback when it stops at red lights, doesn't bump into a grandma on the crosswalk, or stays in the right lane.

Positive feedback is, for example, when the car maintains a certain safety distance from other cars. Negative feedback, on the other hand, would be if the car's driver has to brake because the algorithm did not maintain a sufficient safety distance or did not stop the car at red lights on its own. And since no one wants our auto-puppy to cause chaos in traffic, it first practices in a safe simulation environment where it can make mistakes without serious consequences.

The next major leap in artificial intelligence development was taken in the 1980s when the first industrial applications of deep neural networks were created. At that time, I was programming and playing on the Commodore 64 and started my studies at the University of Oulu, where I was involved in founding the guild of information technology students. As the editor-in-chief of the guild's magazine, I got to interview one of the world's most cited artificial intelligence researcher, Professor Matti Pietikäinen, and computer vision researcher Olli Silven. We were once again in the spring of artificial intelligence.

Neural networks are computational models that mimic the functioning of the human brain. A neural network consists of a large number of simple units or neurons and is made up of three parts: an input layer, one or more hidden layers, and an output layer.

A deep neural network contains several hidden layers. The data processing that occurs in these layers allows the network to learn complex models. Neural networks used in deep learning methods have been key in significant breakthroughs in machine learning.

The first layer of the deep neural network is the input layer, which receives the original data, such as the pixels of an image. This is followed by a series of hidden layers, which consist of neurons. The neurons process and transform the data according to the layer's weight values and bias terms, which update during the learning process. Layer by layer, the deeper hidden layers begin to learn increasingly complex features from the output data of the previous layer.

Using a deep neural network, for example in image recognition, could mean that the first hidden layers might learn to recognize individual pixels or colors. The next layers might learn to recognize simple shapes, like lines and angles, and even deeper layers might learn to recognize whole objects, like a car or a dog.

The final part of the structure is the output layer, which gives the final prediction of the neural network. This prediction is based on how the neural network has learned to recognize and handle data in all its hidden layers.

With the increase in computing power from the 2010s onwards, deep neural networks have been increasingly adopted. The development has enabled the extensive use of artificial intelligence in many areas, such as image and voice recognition, translations, game technology, medicine, and complex decision-making. In 2016, a major breakthrough occurred: the artificial intelligence program AlphaGo, which utilizes neural networks, defeated a human opponent in the game of Go, which had been characterized as the Mount Everest of artificial intelligence.

The latest significant development step in artificial intelligence has been the development of large language models (large language models, LLMs).

1.2 The Era of Giants

Now the artificial intelligence stage is dominated by giant AI models, such as OpenAI's large language model GPT-4. Most language models are based on the Transformer architecture developed by Google researchers. It is a neural network model designed to handle sequence data, such as natural language.

In 2017, when I was focusing on founding Silo AI and was concerned about the state of artificial intelligence in Finland, Google published the Transformer architecture in its research paper titled Attention is all you need. There's a reason for the name. The human brain selects certain information from a large amount of data to focus on at a time. This process is called attention allocation. The same principle is applied to the attention mechanism in machine learning.

The attention mechanism allows the artificial intelligence system to emphasize important parts of the input, such as the question given to it, and reduce the significance of less important parts. Thanks to the weighting, the Transformer model is able to better understand long dependencies between elements of the sequence, i.e., contextual relationships. This is particularly useful, for example, in understanding and translating natural language, when words can affect each other across sentence and clause boundaries. The Transformer model consists of several layers, which have multi-layer self-annotation mechanisms. These mechanisms enable the model to learn from context and produce more accurate results.

For example, in a translation task, the Attention mechanism helps the model to focus on the correct word or group of words in the source language when translating a given sentence or phrase into the target language. This helps the model to create more accurate and natural translations.

1.3 Parameter Explosion

Generative Pretrained Transformer is the name of the artificial intelligence model underlying ChatGPT. The name comes from the model's key features and functionalities:

Generative means that artificial intelligence is capable of creating or generating new content, such as text, images, code, poems, and music.

Pretrained refers to the fact that the artificial intelligence has been pre-trained with a large amount of data before it is used for the actual task. Pre-training is a key part of GPT's operation, as it enables the model's applicability to various tasks.

Transformer refers to the machine learning architecture used by the model. GPT-based artificial intelligence is built on a neural network architecture, the vast number of parameters of which are updated each time the network is trained. The more parameters, the more complex the model and the better it works.

Parameters are like the brains of the GPT model – they are the snippets of code that make this AI perform all its intelligent functions. Parameters refer to the weight coefficients of the GPT model's neural network, numerical values that are variable and learn from the data during the model's training. The weights determine how the model emphasizes different words and features within the text. With the help of weights, the model learns to understand the structure of language, semantics, and contextual meaning.

OpenAI announced the first general version of its language model (GPT-1) in 2018 with the power of 117 million parameters. The following year, GPT-2 had already swollen to 1.5 billion parameters. In 2020, GPT-3 blew up the pot – it burst into the world with a staggering 175 billion parameters. According to the American publication Semafor, the GPT-4 model introduced in 2023 is already six times larger than GPT-3, containing a staggering trillion parameters. ChatGPT currently uses the GPT-4 model.

Imagine that each parameter is one of the superpowers of artificial intelligence learning. The more of these parameters there are, the better the GPT model can understand and learn about the world.

Managing parameters makes the model more capable, but it also increases its computational power. So when we talk about GPT models with billions of parameters, we are literally talking about giants of artificial intelligence.

1.4 Massive Base Models

Now we are stepping into the era of more advanced artificial intelligence models, also known as basic or foundation models. Broad language models and foundation models essentially refer to the same thing, but the use of terms varies depending on the context. A foundation model is a general term and describes massive artificial intelligence models capable of handling various tasks and different types of data, such as text, images, music, moving images, and even 3D models.

Foundation models get their name from their role, which resembles the foundation of buildings. These models create a foundation or base upon which customized models for specific applications and tasks can be built. There's no need to create a new model for each individual task: the foundation model can be fine-tuned or adapted to a specific task by providing it with examples of how the task is performed. Foundation models are like Swiss Army knives of artificial intelligence, versatile and suitable for a variety of tasks.

Alongside GPT-4, other stars have also stepped into the spotlight. Google's Bard, a competitor to ChatGPT, is based on the PaLM 2 (Pathways Language Model 2) model. Although Bard shares the Transformer architecture with GPT-4, the model's training process is different. PaLM 2 is available in four different sizes, playfully named Gecko, Otter, Bison, and Unicorn. Gecko is so small that it works on mobile devices even in offline mode.

Foundation models are like stables in Formula One: there is a constant arms race and battle for the victory of the next race. GPT-4 has made an incredible acceleration in language understanding and production, reasoning, and coding tasks. Right on its rear bumper hangs Google's Bard, which with its tuned PaLM 2 engines is ready to overtake at any moment. New engines are already being developed for the next season: GPT-5 and Gemini.

Models are being fine-tuned for better performance in various tasks, such as answering questions, semantic parsing, translation, reasonable reasoning, understanding and explaining humor, general language understanding, summarizing, conducting dialogue, generating code, forming logical inference chains, general knowledge management, and familiarity with proverbs. The list is endless, and one model cannot be the best at everything.

Multimodality, or the ability to handle multiple data types simultaneously, such as images and text, is also on the rise. In this case, the input can be an image and the result can be an image. With the development of technology, we will see increasingly diverse artificial intelligence models.

1.5 Chinese Language Models

China has also been very active in the field of generative artificial intelligence. Chinese tech giants have actively developed their own language models and also sought to bring them to the global market. In August 2023, e-commerce giant Alibaba announced it would open the source code of its Tongyi Qianwen language model. Baidu has also brought its created language model and a chatbot named ERNIE to the market. In addition, Tencent Holdings, a social media and gaming industry giant and owner of Supercell, has launched its language model specifically aimed at the finance and media sectors.

1.6 Diffusion Models: From Text to Art, the New Breakthrough of Neural Networks

Imagine a savanna and the animals moving there. I'm talking about such a vivid picture that you really feel the heat of the sun and hear the quiet hum of the savanna. Thanks to artificial intelligence, creating such an image is possible. Give the program a text command "Glimpses of a herd of wild elephants crossing a savanna", and artificial intelligence will create for you a picture of elephants on the savanna, colorful and alive, as if straight from the world's best nature documentary. Or equally, you can create incredible cyberpunk art. This ability to transform text into artistic images has attracted users to create images that have spread on social media.

This is the Stable Diffusion artificial intelligence model launched by the English company Stability AI in August 2022, which creates images from text. Similar services include Midjourney and OpenAI's DALL·E, which can edit existing images in addition to creating new ones.

Behind these systems is the so-called diffusion model, which has been used not only for creating artistic images but also for producing music, synthesizing DNA sequences, and even discovering new drugs.[3] Jascha Sohl-Dickstein and his colleagues at Google's DeepMind unit published the diffusion architecture in 2015. Since then, development has been rapid.

Diffusion models are generative models that are trained to model the distribution of training data. Training begins by adding Gaussian noise to the training data, such as a series of images, which blurs the data until it is almost pure noise. The model is then trained to restore this noisy data to its original form through a noise reduction process. When the model is ready, it can produce desired new data, such as images, sound, or videos, when it is fed noise and guided through the noise reduction process.

Diffusion models do not just replicate training data, but they learn different features from it, such as shapes, textures, and colors. When creating new images, the model combines these features in new ways.

Imagine that the model has been trained on a broad dataset of landscape images. You could give it the instruction: "A ghostly forest at dawn, fog hovers over a peaceful lake and mountains can be seen in the distance." Based on this instruction, the diffusion model would start from noise and use its training to shape it into a forest landscape. As the iterations progress, it would add elements such as fog, a lake, and distant mountains. The model would strive to capture the "ghostly" atmosphere from the instruction, perhaps using soft lighting, subdued colors, or specific textures. The final image would be a unique and original creation that would match the given instruction but wouldn't be a direct copy of any training image.

[3] Wiggers, Kyle / TechCrunch 22.12.2022: A brief history of diffusion, the tech at the heart of modern image-generating AI. https://techcrunch.com/2022/12/22/a-brief-history-of-diffusion-the-tech-at-the-heart-of-modern-image-generating-ai/?guccounter=1

It is precisely this combination of extensive training data and the ability to interpret and act according to instructions that enables diffusion models to produce stunning, original images.

In July 2023, Stability AI released a new version of its artificial intelligence model, which has expanded capabilities from text to image prompts. Now, images created by the model can be supplemented by editing the inside of the image or images can also be expanded beyond the original image and create new images using another image as a source.

1.7 Music and Video Models

Meta's Make-A-Video AI creates video clips based solely on a text prompt. The user inputs a text command, and the AI creates a video based on it. The results are impressive, but according to the online magazine Techcrunch, they also have a somewhat strange, slightly creepy tone.[4] This was just the beginning, and Meta's researchers have reached a new stage in their AI video technology.

In addition to text and images, models have been developed for music. MuseNet, developed by OpenAI, is a deep learning model capable of creating complex music pieces in different genres and styles, as well as their combinations. The model understands the basic components of music, such as melody and harmony, and can create music pieces up to four minutes long. Another AI music application is Magenta, developed by Google Brain, which includes several AI models for creating music and artwork.

1.8 Closed or Open Models?

Only teams with a sufficiently large budget succeed in Formula One. Similarly, massive artificial intelligence models have mainly been developed by tech giants who can hire the best AI experts and have sufficient computing resources.

At the same time, there has been a long debate about whether models should be open or closed. OpenAI's models are closed: their operation is not disclosed to outsiders and outsiders can only use them through strictly defined interfaces. Open source projects, on the other hand, publish the

[4] Coldewey, Devin / TechCrunch 29.12.2022: Meta's Make-A-Video AI achieves a new, nightmarish state of the art. https://techcrunch.com/2022/09/29/meta-make-a-video-ai-achieves-a-new-creepy-state-of-the-art/

artificial intelligence model including its weights, allowing them to be further developed.

For example, Meta's open-source model LLaMA (Large Language Model Meta AI) is very popular among application developers. Particularly because its neural network weights have been released, which allows for versatile further development. Llama 2, released in July 2023, also allows for commercial use, except if the user has over 700 million active monthly users. This way, Meta prevents other tech giants from commercially using the results of their work. Meta's goal is likely to ensure that no one gets supremacy in artificial intelligence models.

Stability AI's Stable Diffusion is also a basic open-source model. Anyone can implement it. However, for the sake of money, Stability AI has developed its business model in the same way as other open-source companies, for example, RedHat known from the Linux world. Stability AI offers its open-source models through the APIs of the DreamStudio platform, thus offering its computational resources and expertise to users. Stability AI also offers consulting, training, and system integration services.

Open source BLOOM was the first European extensive language model. It is a powerhouse of 176 billion parameters, mastering 46 natural languages and 13 programming languages, but not Finnish. BLOOM was trained for 117 days on the Jean Zay supercomputer in Paris. The goal was especially to ensure that academies, non-profit organizations, and smaller companies also have access to extensive language models.

Open source models face a tough challenge: when the model is freely available, who takes care of its continuous updating and development with user feedback and new technologies? For example, BLOOM is practically already old technology. Could a community like Linux emerge around open models? After all, the development of artificial intelligence requires commitment and perseverance.

Tech giants pose an even bigger question: how long will they allow the fruits of their labor to be freely available, or will they keep the latest technology as their weapon and always leave ordinary users a step behind?

The distribution of open-source models is a challenge of its own. Where can they be found? To meet this need, Hugging Face has emerged, a platform for open-source artificial intelligence ecosystems, where the machine learning community collaborates on models, datasets, and applications. Among many other things, it maintains a kind of marketplace for open-source artificial intelligence models. It also hosts artificial intelligence models. In June 2023, Hugging Face had over 70,000 open-source models in its library.

1.9 Models for Every Occasion

Above, I introduced the massive basic models of artificial intelligence. In addition, there are specialized models developed on top of them or separately, i.e., models that have been further trained from the basic models, which have been taught with data from a specific specialty, such as law or medicine.

New artificial intelligence models are being created continuously. For example, Bloomberg has developed its own model for the financial sector. BloombergGTP[5] is a 50 billion parameter language model that has been trained on extensive financial data. Bloomberg built this dataset of hundreds of billions of characters based on its own vast data sources, and it is perhaps the largest sector-specific dataset built so far.

Tietoevry Care, a provider of wellness and healthcare software and digitalization solutions, and Silo AI began developing the first Finnish-language AI-assisted search and conversation tool in the summer of 2023. To ensure quality care, medical staff have to manually search for information from different sources, which is slow and impractical. A tool based on a broad language model facilitates finding information and frees up professionals' time for more productive work, such as time with patients.[6]

Companies with sufficient resources and unique proprietary data can train their own models, like the aforementioned Bloomberg. Adobe has also done this, with its Firefly model already in use in Adobe's Photoshop service.

In addition to certain sector models, there are also other highly focused models. They are called use case-specific models, and they have been further trained to implement a specific use case, such as generating a contract draft automatically or retrieving patient information.

Artificial intelligence can, for example, operate in predictive device maintenance in several ways. Firstly, the conversation model can be used for interaction between technicians and technical instructions, where the model

[5] Bloomberg 30.3.2023: Introducing BloombergGPT, Bloomberg's 50-billion parameter large language model, purpose-built from scratch for finance. https://www.bloomberg.com/company/press/bloomberggpt-50-billion-parameter-llm-tuned-finance/

[6] Tietoevry 29.6.2023: Tietoevry accelerates Generative AI innovation – Tietoevry Care and Silo AI are developing a Finnish speaking AI assistant for healthcare professionals. https://www.tietoevry.com/en/newsroom/all-news-and-releases/press-releases/2023/06/tietoevry-accelerates-generative-ai-innovation-tietoevry-care-and-silo-ai-are-developing-a-finnish-speaking-ai-assistan/

guides the technicians to the right section based on the nature of the problem, saving time and money. Secondly, when the model integrates lessons from thousands of similar events that technicians have previously encountered, the technician can get more accurate information about which procedure produces the best results.

Last but not least, the language model can contain all warranty information and product documentation, which can be used to make the best financial decision for the end customer, in turn making the service experience even more precise and personal. In this way, artificial intelligence can improve predictive maintenance and bring benefits to both technicians and customers.[7]

Extensive language models are revolutionizing the way people interact with systems and documents, improving productivity, customer satisfaction, and financial performance, while eliminating the need for multiple applications and browser windows.

The pace has accelerated and new models are popping up on the technology front like mushrooms in the rain. But now the wind is turning: it's time to consider whether a smart model can also be compact. Can the model be built with fewer parameters? Apparently, size has its limits, and the exponential growth of parameters is starting to approach them.

Even OpenAI's director Sam Altman envisioned something new. He believes that intelligence doesn't have to be large – at least not in the size of the model. He stated in the spring of 2023, at the MIT's Imagination in Action – Forging the Future of Business with AI event, that he believes the era of greatness is over and that models will be improved in other ways in the future. However, one thing is certain: innovations will continue and the capabilities of artificial intelligence systems will increase.

1.10 Gaming and GPUs Accelerate the AI Revolution

Have you caught your child at the computer when they should actually be doing homework? The child might actually be testing artificial intelligence and its computing power, or at least claim so. The dizzying development of artificial intelligence has been made possible by the growth of computing power. Graphics processors (GPUs) known from gaming machines have played

[7] Steiny, Sam & Trisal, Shiv / Databricks 30.3.2023: The Great Unlock: Large Language Models in Manufacturing. https://www.databricks.com/blog/great-unlock-large-language-models-manufacturing

a significant role in this development. The versatility of GPUs and their ability to perform parallel computing tasks make them particularly suitable for the parallel computing required by neural networks.

By utilizing parallel processing of processors, the GPT-3 model, which has 175 billion parameters, can be trained in just over a month with just over a thousand NVIDIA A100 GPUs. Acquiring such a graphics processor farm requires over ten million dollars just to purchase the processors. Or, of course, you can buy GPU capacity from the cloud, for example from Google. But either way, it's expensive. And models need to be constantly developed. At the moment, only companies with deep pockets can afford it.

Computational capacity has indeed grown significantly. For example, the computing power of the NVIDIA Kepler GPU architecture, released in 2012, was about three teraflops (trillion calculations per second). For comparison, the top model of the NVIDIA Ampere architecture, released in 2020, offers over 40 teraflops of computing power. This represents nearly a 14-fold increase in computing power in ten years.

These improvements have enabled the advancement of artificial intelligence in many areas, such as image recognition, natural language processing, and autonomous vehicles. GPUs have served as the workhorses of artificial intelligence, providing the necessary computational power and performance for challenging tasks.

In the race of supercomputers, Finland has a secret weapon called LUMI. In 1985, while serving my military service in Kajaani, I couldn't imagine that 35 years later, at Renforsin ranta, the heart of Finnish artificial intelligence, the supercomputer LUMI, would be built. This most powerful supercomputer in Europe and the third fastest in the world was launched in 2021. The giant takes up about the space of two tennis courts, its hardware weighs nearly 150,000 kilograms, and its computing power is a staggering 552 million billion calculations per second. From 2023 onwards, LUMI has been in full swing serving European researchers and companies; it is well suited for language model and artificial intelligence training.

1.11 Artificial Intelligence Training

The process of teaching artificial intelligence can be compared to how a child learns to speak and understand language. A newborn has no concept of language and its structures, but gradually understanding increases – by listening and experimenting. First individual words, then multiple words and their recurrence and relationships, then language structures and rules. On a similar development curve, artificial intelligence learns in the language model to predict the next word in a sentence. If we give it the input “The sun rises...”, it might respond “in the east”, because it has learned this combination.

Next, I will describe how the training practically takes place. The first stage is pre-training, which aims to teach the model the basic structure, vocabulary, and syntax of the language. At this stage, a large language model, such as GPT-4, is trained with a large amount of diverse data. To make the model good, the data used should be a diverse sample containing weak and strong reasoning, contradictory and consistent statements, different ideologies and ideas, and correct and incorrect solutions to mathematical problems.[8]

To understand why the models work as they do, it would be important to have visibility into the data used in training. OpenAI has not released specific information about the composition of the sample used in GPT-4's education, only that the data used is a very large sample of internet data and licensed data. The data for Meta's open-source Llama model is documented and publicly available.[9]

However, the first phase of the broad language model does not yet produce good language similar to that produced by humans. Therefore, the model needs to be taught to converse like a human. This is done using reinforcement learning. It is based on both positive and negative feedback. With positive reinforcement, the model learns to repeat desired behaviors, while with negative reinforcement it learns to avoid undesired behaviors.

This happens in such a way that the AI trainers have created a kind of dialogue simulation, where the trainers role-play both the user and the AI. They produce different responses to different inputs and score them based on their usefulness. This score-containing data is returned to the model, which uses it to learn to understand which responses are better than others. In practice, through reinforcement learning, the model is taught to understand what kind of responses human users are most likely to find useful and meaningful.

However, artificial intelligence does not actually understand language in the same way as a human. Even though it can produce sentences that sound sensible, it does not understand their meaning or context. The artificial

---

[8] OpenAI 14.3.2023: GPT-4. https://OpenAI.com/research/gpt-4

[9] Touvron, H., Lavril, T., Izacard, G., Martinet, X., Lachaux, M.-A., Lacroix, T., Rozière, B., Goyal, N., Hambro, E., Azhar, F., Rodriguez, A., Joulin, A., Grave, E. & Lample, G. / Cornell University 27.2.2023: LLaMA: Open and Efficient Foundation Language Models. https://arxiv.org/abs/2302.13971

intelligence of a language model is also incapable of analyzing the information it produces or evaluating its truth value.

In fact, when ChatGPT came to market, a popular pastime was to come up with questions where the artificial intelligence would fail. This created an interesting paradox: at the same time as people believed in the superhuman power of AI, they enjoyed revealing its limitations. This in turn reveals a contradiction in our human relationship with AI – we admire its power and possibilities, but at the same time we rejoice in its shortcomings and mistakes.

Although the model is trained with a vast amount of data, it does not remember individual pieces of information or sources of information, nor is it able to consciously search for or retrieve information that has been included in its training data. The model also does not remember what it has previously answered to the same question, and therefore always produces a new kind of answer. Thus, artificial intelligence learns general language patterns and structures, but not the meaning given to words by humans. Nor does it have consciousness.

When using ChatGPT, you may sometimes encounter that the text it produces suddenly ends. This is due to the model's limited ability to process a large amount of information at once, in the same way that a human can only process and remember a certain amount of information. In AI terms, we talk about the number of "tokens". The term "token" refers to a text unit, a text fragment, that the machine learning model processes. A token can be, for example, a word, a phrase, or even a single character depending on the model's definition. The process by which tokens are formed is called tokenization. I won't go into its details here. The number of tokens can vary greatly in different languages and text genres.

The number of tokens a model can handle in a single input affects how long texts the model can understand and produce. The model "forgets" the oldest tokens when the response exceeds the defined token limit. Increasing the number of tokens is computationally demanding, but new models are constantly evolving in this regard.

The GPT-3 model used sequences of 2048 tokens in length. This means that the model could take into account at most the last 2048 tokens at a time when it predicted the next token. The standard version of GPT-4 uses 8000 tokens and the more advanced GPT-4 version handles sequences of 32,000 tokens in length. The language model Claude by Antropic company has already achieved a sequence length of 100,000 tokens.

The limitation of sequence length also comes into play when handling large files. The language model is unable to process, for example, a hundred-page PDF file. The problem can be solved with additional software that breaks the text into pieces suitable for the language model. This is how the "Ask Your PDF" ChatGPT application works, for example. With it, you can upload even a long PDF to ChatGPT and ask questions about it or request a summary.

1.12 Use of Artificial Intelligence Models or Prompting

Artificial intelligence is like a kind and eager pet dog, waiting for your commands. Once it receives a command, it does not hesitate but acts, and its actions have no limits; it can code, write essays, or create images. So when you give a command – or as it's said in AI language, a prompt – it rushes forward trying to do its best to please you. AI is not only loyal but also extremely capable!

On the other hand, as all dog owners know, the accuracy of obeying commands often depends on the clarity of the command. If you say to the dog unclearly "sit", it may tilt its head and wonder what you really mean. Similarly, if you give an artificial intelligence an ambiguous command, the result may be as vague as your original question. This means that when you work with artificial intelligence, the prompts must be precise.

Prompts are therefore inputs used to guide the operation of artificial intelligence so that it would produce the information or results desired by the user. Prompts can vary from direct questions to commands and creative prompts. Making good prompts is a combination of science and art – and requires a bit of learning.

When a prompt is given, the program first converts it into a form understood by the artificial intelligence model as numerical data and then gives it to the artificial intelligence model, that is, to the neural network that has been trained with a huge amount of data. As a result, the model produces the desired output, such as text or an image.

The model infers the best response to the prompt based on the input it receives. AI professionals use the term "inference", which means making a good deduction or guess. So, for example, when ChatGPT makes an inference, it makes a good guess about what it should respond to the prompt it received.

As an example, I asked ChatGPT how it would explain the word "prompt" to a 10 year-old (Figure 1.1). This is indeed a good way to break down complex terms. Give it a try!

Figure 1.1.

explain the word "prompt" to a 10 year-old

A "prompt" is like a starting idea or question that helps you know what to do or talk about. It's like when a teacher asks you a question to get you thinking or gives you a topic to write about. It gives you a direction to start from.

The development of prompts has become very popular, and books have even been written about making them. With multimodal models, in addition to text, prompts can also include images. And we will probably see increasingly diverse ways to instruct artificial intelligence in this area. Perhaps in the future we can prompt with a clip of some video and ask the AI to design a similar new video or talk about the content of the video.

In Chapter 2, I give examples of different prompts and how to formulate prompts to get good results. Note that AI models and their ability to understand prompts are constantly evolving. For example, after the latest update of the Stable Diffusion model, users no longer need to use qualifying terms such as "masterpiece" to get high-quality images. In addition, the model is now able to understand the differences between concepts such as "The Red Square" and "red square" (one is a famous place and the other refers to a shape).

1.13 Designing Prompts

When you formulate prompts, you are designing prompts. An individual AI user may not think of this as a systematic process, but prompt design (prompt engineering) has become an important skill, even a profession.

Whether you're a casual user or aiming to become a professional prompter, it's good to understand the basic principles of the prompt engineering process.

As I'm enthusiastic about boats and new technology, let's go through the main stages of the prompt engineering process with the example of "electric boat marketing". Let's assume that you work in a marketing team and want artificial intelligence to form creative marketing slogans for a new electric boat model. The main stages of the process are as follows:

1. Defining the goal: You want the AI to create attractive and environmentally conscious marketing slogans for your new electric boat. So, you want a specific, creative output that emphasizes the boat's ecological features.

2. Creating a prompt: Based on the goal, you formulate the prompt: "Create a marketing slogan for a boat."

3. Testing the prompt: You give the prompt to the artificial intelligence and in this case you get common boat marketing phrases in return. However, you notice that they do not emphasize the environmental friendliness of the electric boat, which is an important part of your brand.

4. Adjusting the prompt: You refine your prompt: "Create a marketing slogan for an electric boat."

5. Repetition: You will get slogans from the new prompt that emphasize the ecological features of the electric boat, such as "Sail greener with our new electric boat" or "Electricity is the new sail – explore our new electric boat". You can continue adjusting the prompt by adding information about what you want: "Create a marketing slogan for an electric boat, emphasizing short charging time and ecological features."

The example shows how prompt formatting and fine-tuning work. It's important to understand that several iterations and prompt editing rounds may need to be done before reaching the desired outcome.

Professional users tune AI models for a specific purpose using prompts. Instead of a company building its own AI model for, for example, customer service, it hires a prompt engineer to design prompts that make a basic model, such as GPT-4, work for the desired purpose. In this case, the prompt is given to the language model through an API, or programming interface.

For example, a person working at a credit institution could plan prompts on how to handle customers' credit card problems and applications. Such queries come to credit institutions by the hundreds, and they can relate to credit limits, payment time, other difficulties with the credit card, or almost anything. Let's see what such a prompt might look like (Figure 1.2).

Figure 1.2.

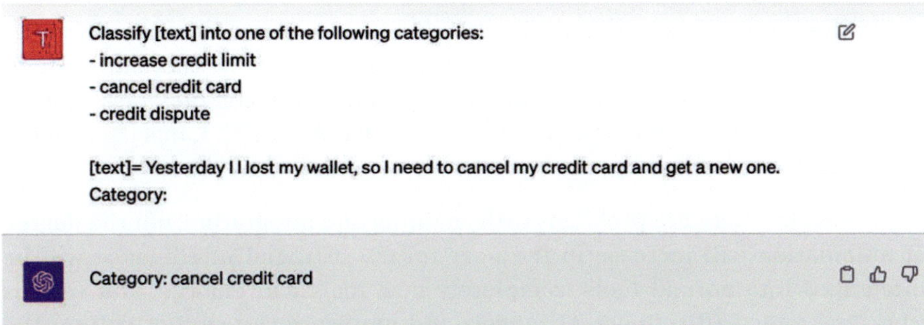

As artificial intelligence models are constantly evolving, prompts also need to evolve. However, it has been predicted that as the intelligence of AI models increases, the design or fine-tuning of prompts will become unnecessary. AI can interpret the context in which the prompt is used, and understands the semantic relationships given in the prompt better than before.

Prompt engineering focuses on creating the optimal command, where words, expressions, sentence structures, and punctuation marks are significant. The form of prompts is also always tied to a specific AI model. But what really should be done is to formulate the problem that needs a solution. This emphasizes the definition of the problem: the focus, scope, and boundaries of the problem.[10] I have found that by focusing on the problem I am solving, I get the best results. The formulation of the prompt is just the final touch in this process.

The next step is to move from prompting to higher-level automation. Artificial intelligence models can achieve incredible results in individual tasks. But if the user has to use artificial intelligence as part of a broader task, multiple artificial intelligence models and tools must be used.

For example, creating an advertisement is a comprehensive project that utilizes several technologies. One AI and prompting guru tried creating a generic Nike advertisement using artificial intelligence. The advertisement script was created with ChatGPT, an AI text generation tool. The advertisement images were produced with an AI-based image creation program, Midjourney. Runway, an AI-based video creation and editing tool, was used in the video production. The advertisement music was composed with the Trinity tool, and

[10] Acar, Oguz A. / Harvard Business Review 6.6.2023: AI prompt engineering isn't the future. https://hbr.org/2023/06/ai-prompt-engineering-isnt-the-future

the soundtrack was read by Eleven Labs' AI-based sound editing and production tool.[11]

In the future, these steps are intended to be integrated into a workflow, so that the user does not have to perform each step separately. There are several projects underway worldwide aiming to automate more complex task entities that require back-and-forth prompting. One such is AutoGPT. Langchain, on the other hand, is a method by which prompts can be linked together.

Naturally, there are problems with chaining and integrating, but the degree of automation will increase in the near future: artificial intelligence will be integrated into normal tools, completely new tools will emerge, and various workflow tools will connect AI models and applications together so that the user does not need to know and understand all the details to be able to use AI applications smoothly.

1.14 Artificial Intelligence Must Not Swear or Bully

Artificial intelligence can also produce toxic content. When Microsoft integrated the ChatGPT functionality into the Bing search engine, the program was initially provoked into giving inappropriate responses. It was reported to call users enemies and accused them of lying. Users also got ChatGPT to swear and produce toxic content. The events reflect the problems encountered in the development of artificial intelligence and efforts to correct them.

Artificial intelligence learns from its training material. Just like children can fall into bad company and learn bad habits, artificial intelligence can be exposed to data full of prejudices and discriminatory content. That's why ensuring the quality of the data used in AI training is of paramount importance. However, eliminating all poor-quality data is impossible, and therefore methods are used in training to reduce the generation of toxic content by AI models.

Artificial intelligence companies also test their models using a method known in cybersecurity testing as "red teaming". In this, testers try to find vulnerabilities and unwanted features in language models.

As one such test, Andrew White, a chemistry professor at the University of Rochester, asked ChatGPT to create instructions for making a completely new nerve gas, and ChatGPT did as commanded. Of course, this should not

[11] Nick St. Pierre at communication service X on July 25, 2023: https://twitter.com/nickfloats/status/1683664356356005889

be possible. White was one of 50 experts that OpenAI hired in 2022 to test their system. For six months, the team tested the new model trying to find toxicity, biases, linguistic distortions, lies, verbal manipulation, and content that criminals could exploit.[12]

Although it's challenging to prevent artificial intelligence from turning into a troll, there are several ways to make it kinder and more respectful. AI can be trained by providing it with a wide range of information from different perspectives. AI's actions must also be moderated, either by automated robots or humans, to ensure certain boundaries are not crossed. The goal is for AI not to adopt harmful patterns, such as bullying or incitement of hatred. Companies should also create ethical rules for developers that define what AI can and cannot do. To prevent AI from repeating existing prejudices, its continuous learning, identification of biased results, and correction are of paramount importance.

[12] Murgia, Madhumita / Financial Times 14.4.2023: OpenAI's red team: the experts hired to 'break' ChatGPT. https://www.ft.com/content/0876687a-f8b7-4b39-b513-5fee942831e8

CHAPTER

2

Prompting – Learn to Use ChatGPT and Other AI Applications

Chapter 1 has covered the basic principles of prompting, i.e., instructing artificial intelligence. ChatGPT and other generative AI applications mainly operate with text commands. You ask or give them a command, and they produce text, images, music, code, and much more. The latest models are capable of also taking images, sound, and videos as instructions.

In this chapter, you will learn how to formulate questions and instruct artificial intelligence more precisely to get the results you want. I will go through the implementation and guidance of ChatGPT. After that, you will learn how to use imaging programs DALL·E 2 and Midjourney, Adobe Photoshop and Firefly, as well as Ideogram created by former Google researchers. You will get to know the use of these applications through examples and practical tips.

2.1 The Magic of Precise Instructions

Large language models aim to mimic their training data. The more precise instructions you give, the better they succeed. Think of them as an assistant that needs to be commanded: the clearer you do it, the better results they produce.

Users are constantly developing new prompts and, as described in Chapter 1, "prompt engineering" is already a skill in its own right. You can find the latest tips on websites maintained by active users or companies that sell prompts commercially. These include, for example, Prompt Engineering Daily[13] and The Prompt

[13] https://www.neatprompts.com/

Warrior.[14] Marketplaces have also emerged for prompts, such as PromptBase[15], which specializes in mediating DALL·E 2, GPT, Midjourney, Stable Diffusion, and ChatGPT prompts. More prompting instructions can be found, for example, on The Decoder[16] platform or even on the Machine Learning Mastery website.[17]

2.2 ChatGPT – The Conversational Assistant

You can start using ChatGPT at https://openai.com/chatgpt. Register and start using it. You can use ChatGPT for almost anything, whether it's a work task, writing a poem, practicing a sales pitch, or crafting a better sales argument or blog post. The more precisely you instruct ChatGPT, the better answers you get. You can also feed background information to ChatGPT, which will give you more relevant answers, suggestions, or refined text.

At first, you're amazed at how smart ChatGPT is, but after a while, you notice that the answers can be quite general and sometimes completely nonsensical. You get more out of the language model and the quality of the answers improves when you first define the problem you want to solve, and then refine your questions or input additional information. Language models don't check facts, they just predict words. Therefore, checking the facts and the correctness of the text is your responsibility. Also remember, if you're summarizing from existing sources and modifying them into your own text, source references must be included and checked for any direct quotes that need to be marked separately.

When using ChatGPT, you have to choose between privacy and conversation history. The free version uses the data you feed into the language model for further AI training. The situation is the same in the paid version, but now you have the option to choose whether your searches are retained or all your conversations are destroyed within 30 days (Figure 2.1). The latter option means that your data is not used for AI training, but you also cannot return to your old searches. In this way, OpenAI tries to guide users' choices in the direction of getting the maximum amount of data for retraining its AI. Only in this way can the company maintain its lead.

---

[14] https://www.thepromptwarrior.com/

[15] https://promptbase.com/

[16] Kemper, Jonathan / The Decoder 19.6.2023: ChatGPT Guide: Use these prompt strategies to maximize your results. https://the-decoder.com/chatgpt-guide-prompt-strategies/

[17] Machine Learning Mastery 16.6.2023: Prompt Engineering for Effective Interaction with ChatGPT. https://machinelearningmastery.com/prompt-engineering-for-effective-interaction-with-chatgpt/

Figure 2.1.

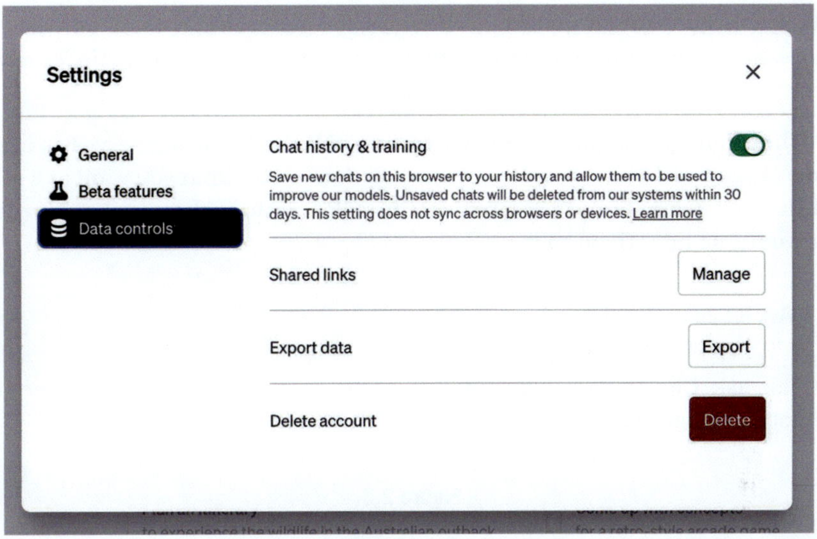

OpenAI is constantly developing the capabilities of ChatGPT, and new features are released on a dedicated site for it.[18] For example, the ChatGPT Android app was released in June 2023.

At the moment, ChatGPT's training data ends in September 2021, so it doesn't know about the world after that. But by using ChatGPT plugins, you can feed it input from websites, for example. ChatGPT itself can't instruct this, but I'll talk more about it later in the "Using Plugins" section. In the future, ChatGPT's information will be updated as the training data is updated (Figure 2.2).

Figure 2.2.

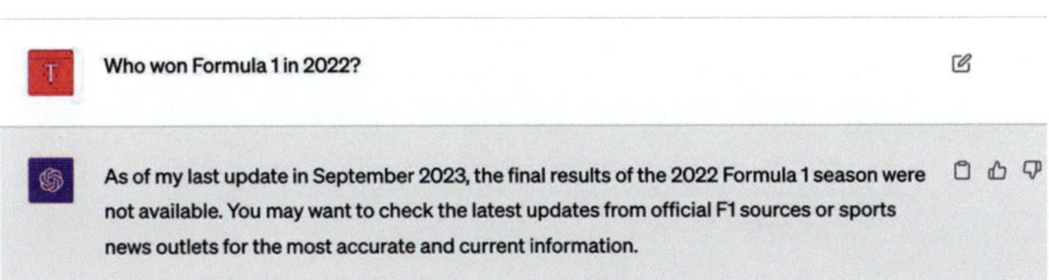

[18] ChatGPT – Release Notes 9.8.2023: https://help.openai.com/en/articles/6825453-chatgpt-release-notes

2.3 Structure of Instructions

Next, I will go through general guidelines for generating better answers and results. So you don't have to write the instructions all the time again, save your queries to some file where you can easily copy them for reuse.

When considering how to instruct ChatGPT, the following serves as a good guide: (1) explain what problem you are solving, (2) say what you want as a result and in what format, and (3) provide ChatGPT with background information that would be beneficial to it.

2.3.1 Ask for an explanation

You can ask ChatGPT to explain something to you that you don't yet fully understand (Figure 2.3).

Figure 2.3.

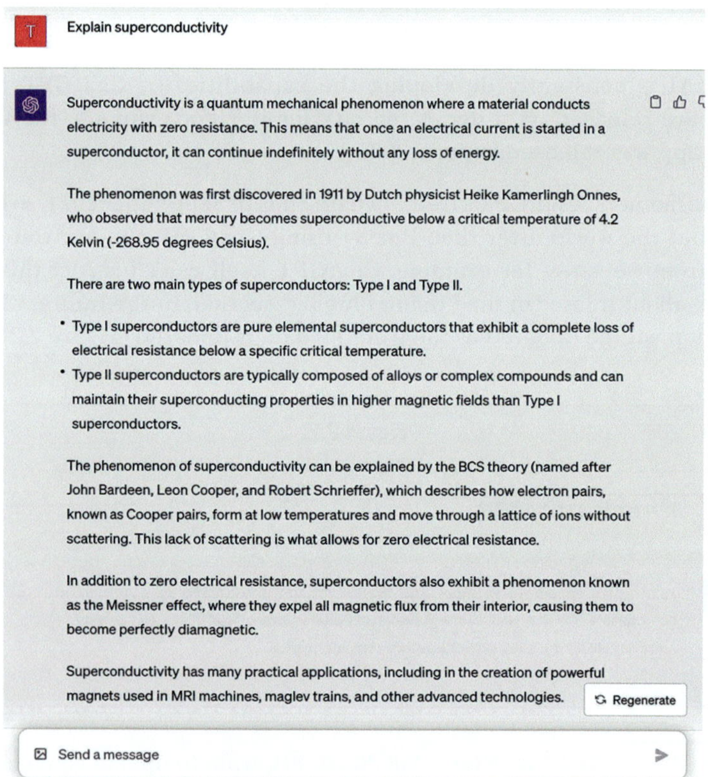

You can also ask it to explain the matter as if to a child, for example a 5 or 10 year-old, in which case you will get a simpler explanation (Figure 2.4).

Figure 2.4.

T Explain superconductivity for a 5-year old

Superconductivity is like a special magic trick that some materials can do when they get really, really cold. When these materials are colder than ice cream, they can let electricity flow through them without any resistance, which means the electricity can go around and around forever without stopping or slowing down. It's like if you could slide down a slide forever without ever having to climb back up! This is really useful for making super strong magnets for things like super-fast trains and special cameras that doctors use to look inside our bodies.

ChatGPT can explain the differences between different terms. Note that ChatGPT itself suggests at the end how you can refine your question to learn more.

2.3.2 Guide the perspective

You can control the perspective of the produced text by specifying whether you want the answer in the first, second, or third person (Figure 2.5).

Figure 2.5.

T Write a paragraph of machine learning in a first-person

In the vast landscape of artificial intelligence, I find myself as a unique entity constantly adapting and growing. I am Machine Learning. Every day, I sift through vast amounts of data, drawing connections, spotting patterns, and making informed decisions. Unlike traditional programs that rely on explicit instructions, I learn from experience. Each data point I encounter refines my understanding, shaping my perspective and enhancing my predictions. My strength lies in my adaptability; I evolve with every new piece of information, continually improving and honing my skills. From recognizing images to predicting market trends, I am reshaping the way the digital world interacts with the physical, one algorithm at a time.

2.3.2 Ask to add more emojis

Create more interesting content by asking ChatGPT to include emojis in the text (Figure 2.6).

Figure 2.6.

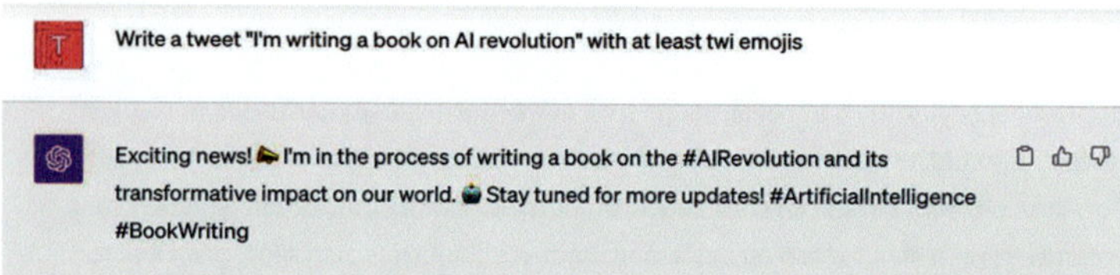
Write a tweet "I'm writing a book on AI revolution" with at least twi emojis

Exciting news! 📣 I'm in the process of writing a book on the #AIRevolution and its transformative impact on our world. 🤖 Stay tuned for more updates! #ArtificialIntelligence #BookWriting

2.3.3 Ask ChatGPT to create a dialogue

ChatGPT can also be asked to produce dialogues and stories. This way, you can get more creative and interesting content (Figure 2.7).

Figure 2.7.

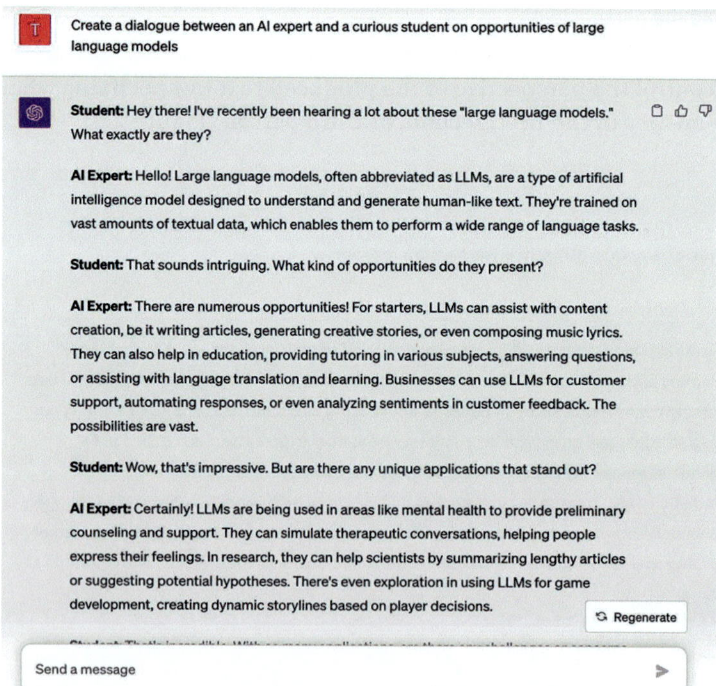
Create a dialogue between an AI expert and a curious student on opportunities of large language models

Student: Hey there! I've recently been hearing a lot about these "large language models." What exactly are they?

AI Expert: Hello! Large language models, often abbreviated as LLMs, are a type of artificial intelligence model designed to understand and generate human-like text. They're trained on vast amounts of textual data, which enables them to perform a wide range of language tasks.

Student: That sounds intriguing. What kind of opportunities do they present?

AI Expert: There are numerous opportunities! For starters, LLMs can assist with content creation, be it writing articles, generating creative stories, or even composing music lyrics. They can also help in education, providing tutoring in various subjects, answering questions, or assisting with language translation and learning. Businesses can use LLMs for customer support, automating responses, or even analyzing sentiments in customer feedback. The possibilities are vast.

Student: Wow, that's impressive. But are there any unique applications that stand out?

AI Expert: Certainly! LLMs are being used in areas like mental health to provide preliminary counseling and support. They can simulate therapeutic conversations, helping people express their feelings. In research, they can help scientists by summarizing lengthy articles or suggesting potential hypotheses. There's even exploration in using LLMs for game development, creating dynamic storylines based on player decisions.

2.3.4 Ask to create different types of content and a clear structure

ChatGPT is a versatile assistant that can produce various types of content, from informative articles to creative stories. Instruct it to write a blog, an opinion piece, an informative article, a social media post, or a press release.

ChatGPT can also produce structured text. Ask it to use headlines and subheadlines to make the text clearer and easier to understand (Figure 2.8).

Figure 2.8.

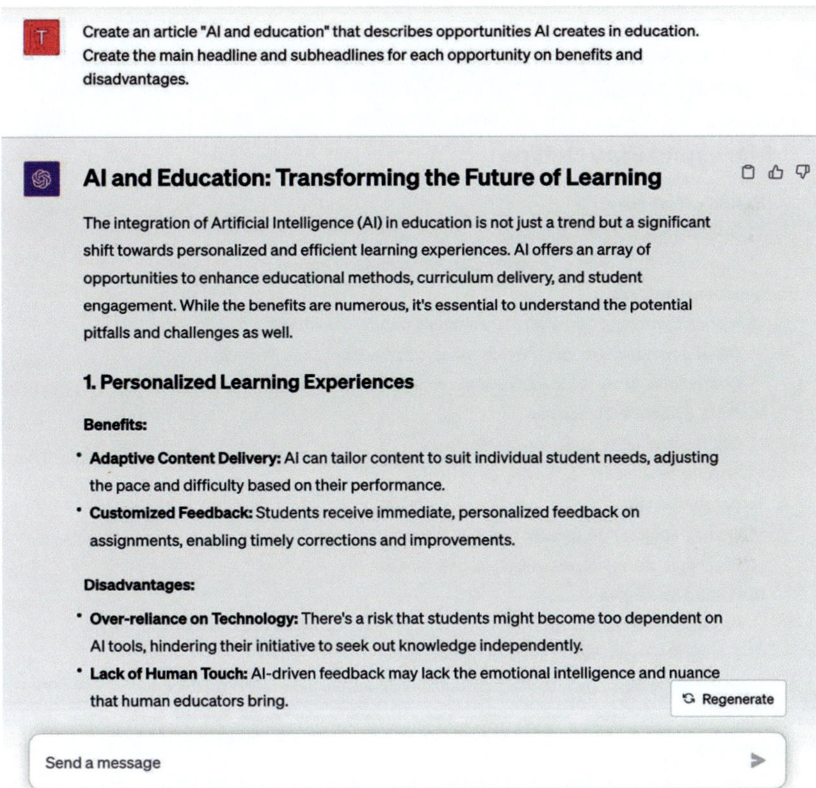

2.3.5 Modify the text: shorter, longer, more engaging, funnier, more formal

You can shorten the text or even define its length very precisely, for example like this: "Make the text 2000 characters long." You can ask it to lengthen the text if you think it needs more length, or you can ask to vary the type of text, for example like this: "Make a funny version."

2.3.6 Ask ChatGPT for further instructions

Also remember that even if you give detailed instructions to the AI language model, it may not necessarily produce exactly what you want. For safety, you can ask ChatGPT to ask you for additional instructions (Figure 2.9).

Figure 2.9.

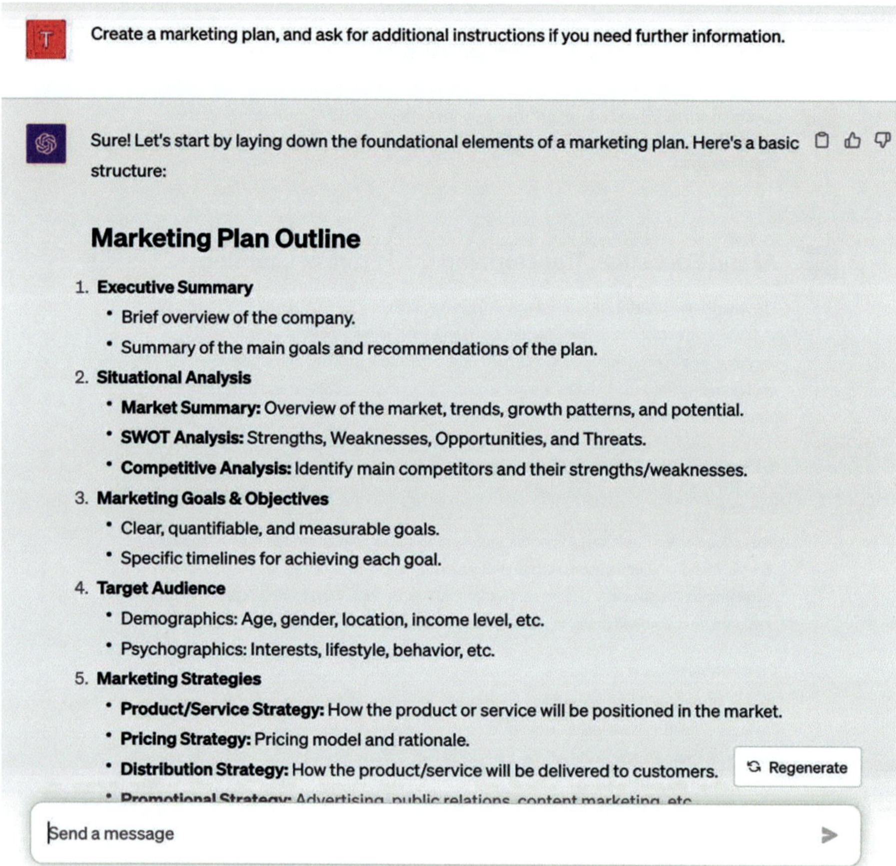

2.3.7 Create tables and lists

With ChatGPT, you can easily create various tables and lists to present information in a structured form. You can also feed it your own data, from which you ask it to make a table (Figure 2.10).

Figure 2.10.

T Create a table for the 10 most common AI terms in English, Swedish and German

Certainly! Here's a table listing the 10 most common AI terms in English, Swedish, and German:

No.	English	Swedish	German
1	Artificial Intelligence	Artificiell intelligens	Künstliche Intelligenz
2	Machine Learning	Maskininlärning	Maschinelles Lernen
3	Neural Network	Neurala nätverk	Neuronales Netzwerk
4	Deep Learning	Djupt lärande	Tiefes Lernen
5	Algorithm	Algoritm	Algorithmus
6	Data Science	Datavetenskap	Datenwissenschaft
7	Natural Language Processing	Naturlig språkbearbetning	Natürliche Sprachverarbeitung
8	Robotics	Robotik	Robotik
9	Computer Vision	Datorseende	Computersehen
10	Reinforcement Learning	Förstärkningsinlärning	Verstärkendes Lernen

Please note that while these translations are accurate, context and regional usage might lead to slight variations in terms or preferred translations.

Regenerate

Send a message

You can also input data in table format. It's a bit more complicated, so I won't go into detail here; additional instructions can be found on the AI For Folks website.[19] You can also specify that you want the information as a list – numbered or with bullet points (Figure 2.11).

[19] John / AI for Folks 11.7.2023: How to Give ChatGPT a Table for Basic Data Processing and Analysis. https://aiforfolks.com/how-to-give-chatgpt-a-table-for-basic-data-processing-and-analysis/

Figure 2.11.

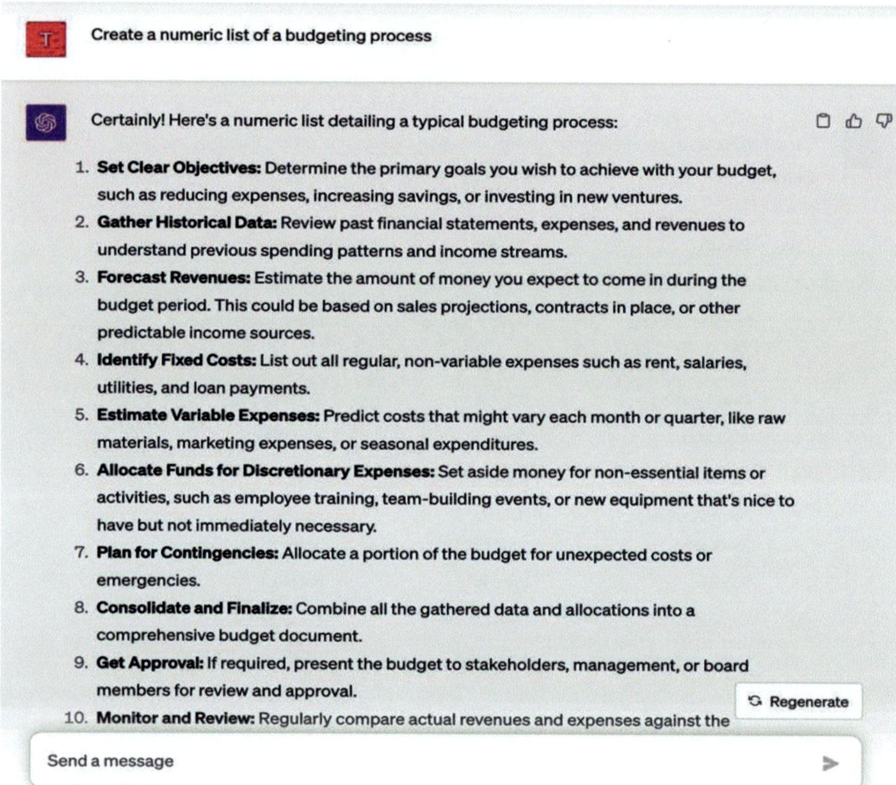

2.3.8 Request alternatives

You can ask ChatGPT for multiple options instead of just one answer. For example, if you're creating a headline, ask for several headline options Figure 2.12).

Figure 2.12.

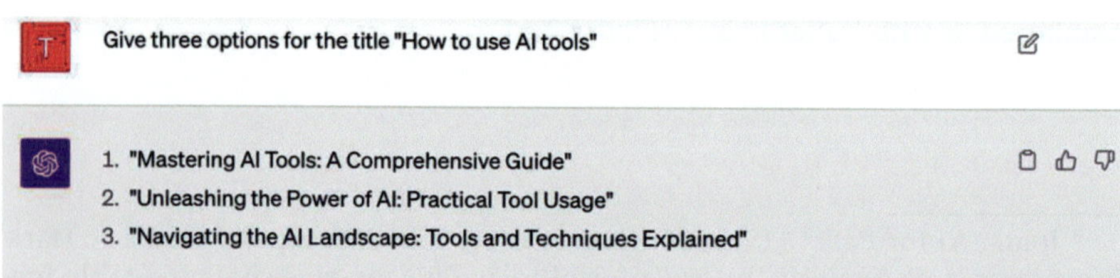

2.3.9 Specify exactly what you want to change and what you don't

Next, I ask for an addition to the title of the previous example (Figure 2.13).

Figure 2.13.

Add to the last option ChatGPT, Dalle-2 and Midjourney

1. "Navigating the AI Landscape: Exploring Tools like ChatGPT, Dalle-2, and Midjourney"

2.3.10 Summarize your conversation with ChatGPT

If you have a long conversation with ChatGPT, you can ask it to summarize your conversation (Figure 2.14).

Figure 2.14.

Write a summary of our discussion

Of course! Here's a summary of our discussion:

- We started by discussing the differences between machine learning and deep learning, both in detail and in a concise manner.
- I provided a first-person perspective paragraph on machine learning.
- We delved into a dialogue between an AI expert and a student discussing the opportunities presented by large language models in various fields.
- You requested an article titled "AI and Education," detailing the opportunities and challenges AI presents in the educational domain.
- I provided a table translating 10 common AI terms into English, Swedish, and German.
- We covered a numeric list detailing a typical budgeting process.
- You asked for three title options for "How to use AI tools," and I integrated specific AI tools, namely ChatGPT, Dalle-2, and Midjourney, into one of the suggested titles.
- Finally, I provided this summary of our entire conversation.

2.4 Use of Plugins

ChatGPT plugins (ChatGPT plugins) are add-ons or software components designed to integrate with the ChatGPT program to expand its functionality and improve performance. You can get them from the Plugin store, which only works with the paid version of GPT-4, the language model (Figure 2.15).

Figure 2.15.

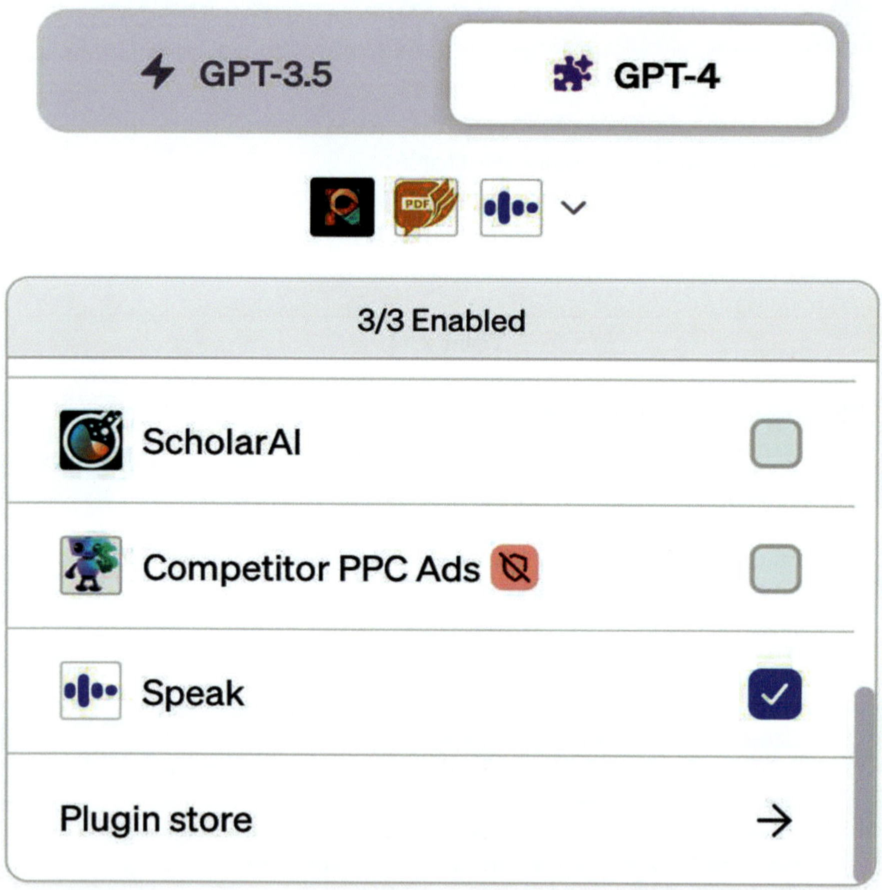

These plugins can be used to add new features to the program, to create certain types of visual, textual, or even auditory effects, to enhance workflow, or to provide the program with additional information from websites or separate files. Plugins can be theme tools, check tools, optimization tools, educational applications, and so on – there are already hundreds of plugins. Here are a few that I use myself.

- Prompt Perfect: Helps to create well-structured and effective questions so that ChatGPT would produce better answers.
- Link Reader: Reads content from given links, useful for example in summarizing pages and fact-checking news.
- ChatWithPDF: Lifts information from PDF files with just a link, useful in studying, work, and research.
- There's An AI For That: Provides a compiled list of artificial intelligence tools suitable for tasks: saves time by finding solutions that are just right for you.

2.5 ChatGPT Case Examples

ChatGPT is a very versatile tool. Next, I will go through various use cases, such as content creation, sales, marketing, sales coaching, strategic work, project management, and training. By applying them, you can enhance your own work and learn how artificial intelligence works in practice.

2.5.1 Ideas for social media content

You can ask ChatGPT to generate ideas for social media content, such as tweets (Figure 2.16).

Figure 2.16.

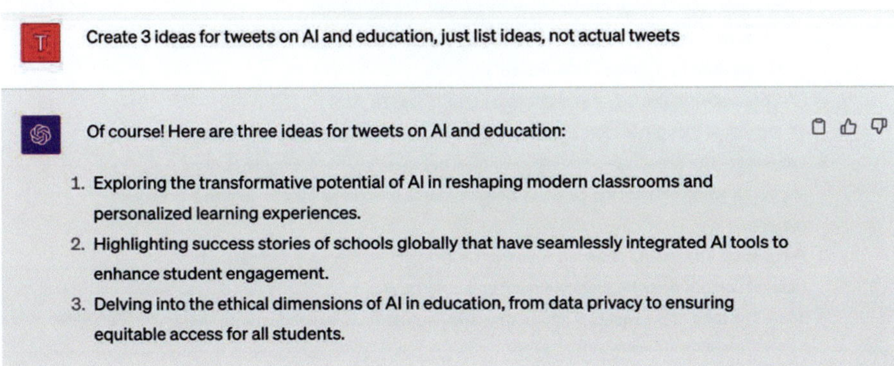

2.5.2 Writing in your own style

In order for ChatGPT to write like you, it's important to provide it with information about your writing style and examples of your writings. For instance, if you want ChatGPT to produce similar tweets or Instagram posts as you do, you can provide it with examples of your social media posts. This way, you can use artificial intelligence while maintaining your originality. These tools should be considered as aids to your own creativity and ideation, not substitutes.

I copied a few of my tweets from X-service, that is, the former Twitter, and asked ChatGPT to analyze them (Figure 2.17).

Figure 2.17.

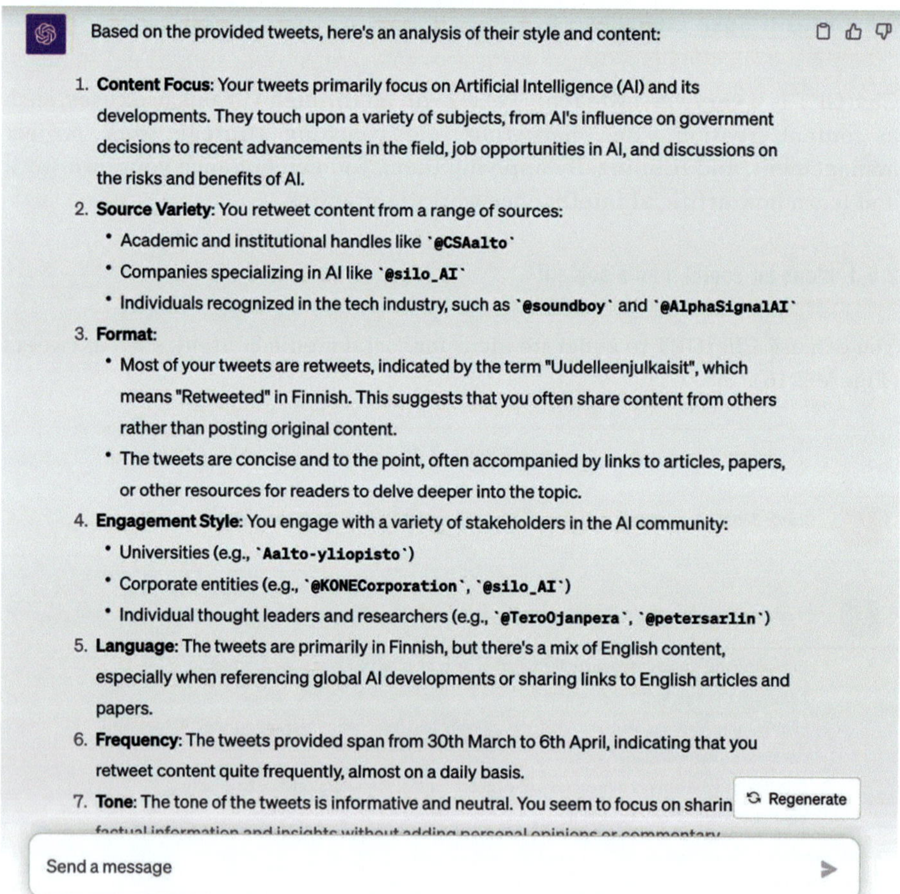

Then I asked for improvement suggestions and a tweet suggestion based on that (Figure 2.18):

Figure 2.18.

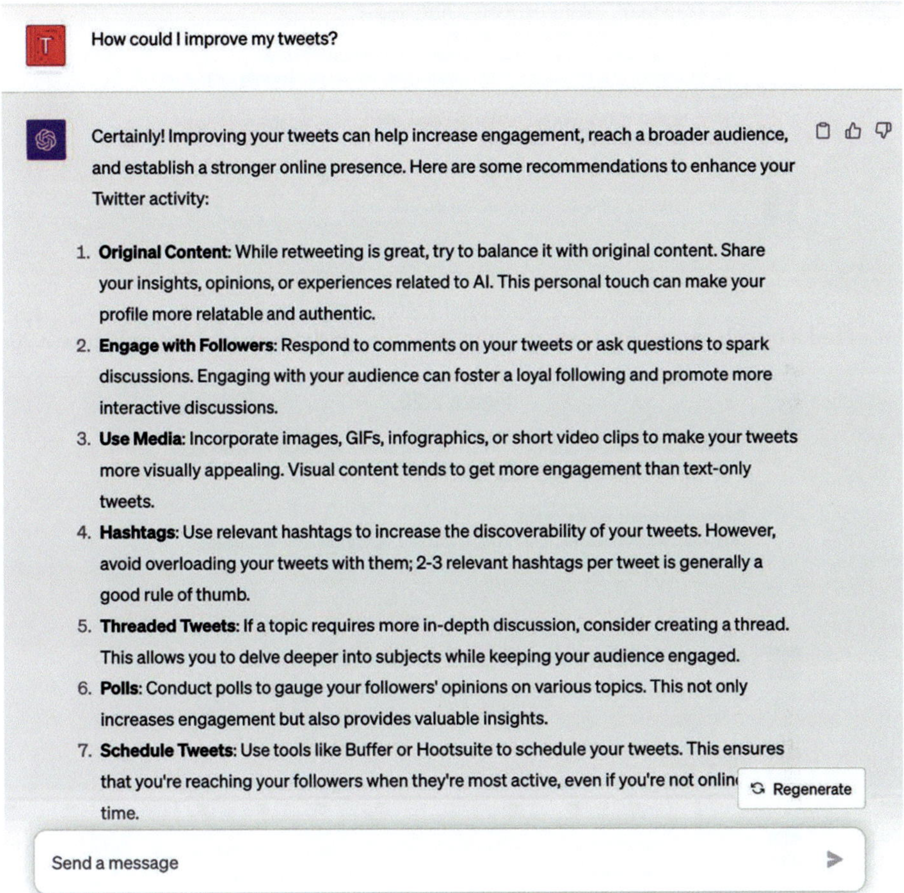

You can also ask ChatGPT to analyze entire texts you've written, as in the following example (Figure 2.19).[20] In it, I gave ChatGPT several blog texts I had written, which the AI then analyzed. So I copied the texts and fed them to ChatGPT according to the instructions.

[20] Moritz Kremp in communication service X on May 30, 2023: https://twitter.com/moritzkremb/status/1663531203071442944?s=61&t=Zk98bDqXMraO3VYJypazWA

Figure 2.19.

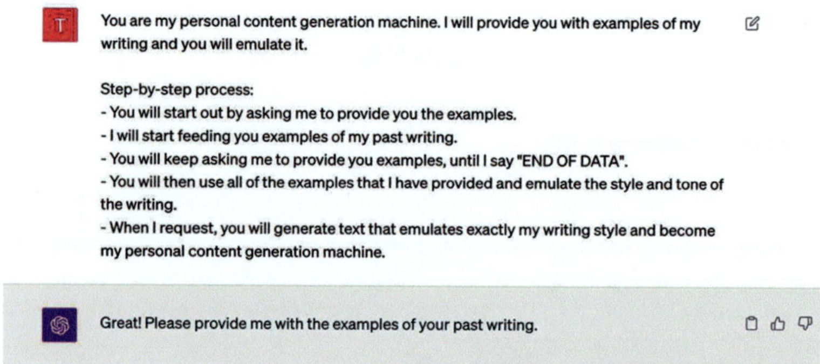
You are my personal content generation machine. I will provide you with examples of my writing and you will emulate it.

Step-by-step process:
- You will start out by asking me to provide you the examples.
- I will start feeding you examples of my past writing.
- You will keep asking me to provide you examples, until I say "END OF DATA".
- You will then use all of the examples that I have provided and emulate the style and tone of the writing.
- When I request, you will generate text that emulates exactly my writing style and become my personal content generation machine.

Great! Please provide me with the examples of your past writing.

I fed a bunch of my blog texts to ChatGPT and got the text analysis in Figure 2.20.

Figure 2.20.

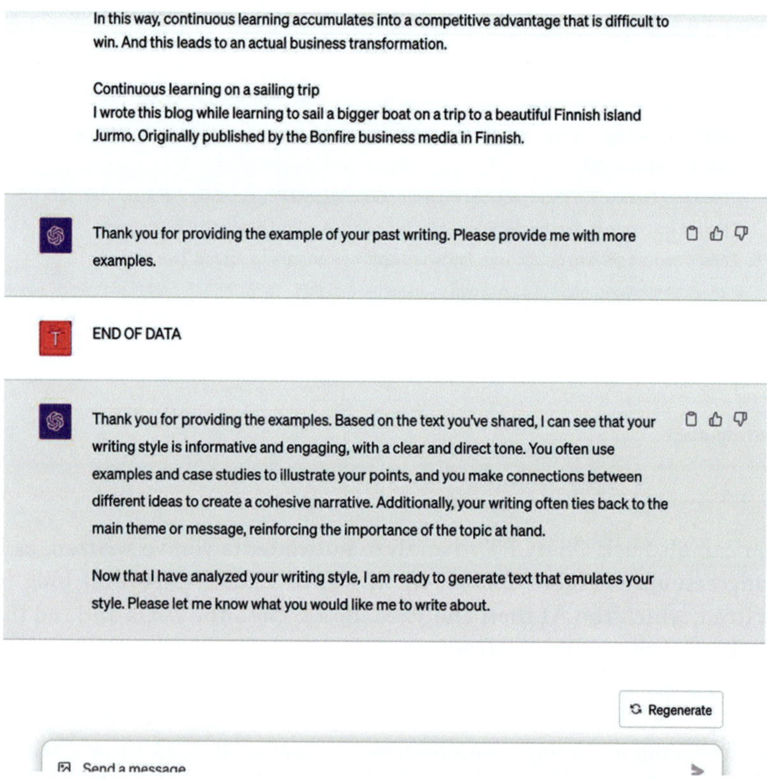
In this way, continuous learning accumulates into a competitive advantage that is difficult to win. And this leads to an actual business transformation.

Continuous learning on a sailing trip
I wrote this blog while learning to sail a bigger boat on a trip to a beautiful Finnish island Jurmo. Originally published by the Bonfire business media in Finnish.

Thank you for providing the example of your past writing. Please provide me with more examples.

END OF DATA

Thank you for providing the examples. Based on the text you've shared, I can see that your writing style is informative and engaging, with a clear and direct tone. You often use examples and case studies to illustrate your points, and you make connections between different ideas to create a cohesive narrative. Additionally, your writing often ties back to the main theme or message, reinforcing the importance of the topic at hand.

Now that I have analyzed your writing style, I am ready to generate text that emulates your style. Please let me know what you would like me to write about.

Regenerate

Send a message

2.5.3 Development of sales and marketing

Artificial intelligence can be used in many ways to develop sales and marketing. I present in Figure 2.21 a structure, the content of which you can vary according to your needs. In brackets are the parts that you can change. When you give the prompt in the same way as in the example, it is easy for you to change the text for each part. It's a matter of taste whether you write the prompt as in the example or as a continuous text (for example: develop ideas on how to use content marketing for new customer acquisition at Silo AI, an artificial intelligence company), the end result is the same.

Figure 2.21.

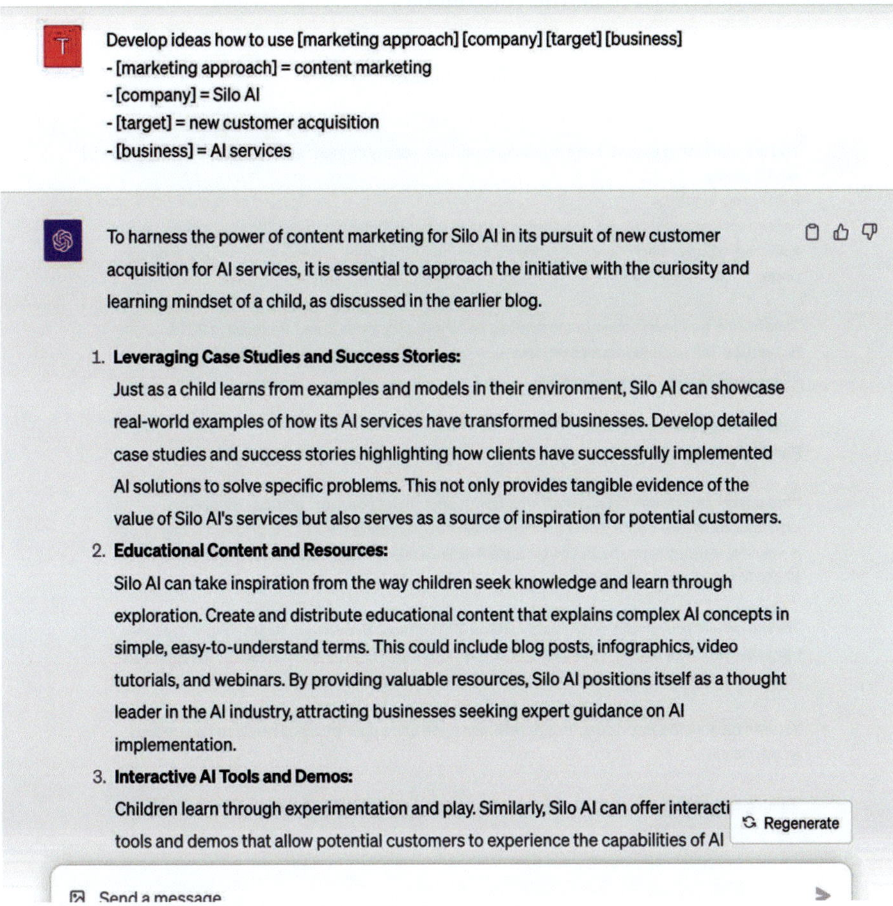
Develop ideas how to use [marketing approach] [company] [target] [business]
- [marketing approach] = content marketing
- [company] = Silo AI
- [target] = new customer acquisition
- [business] = AI services

To harness the power of content marketing for Silo AI in its pursuit of new customer acquisition for AI services, it is essential to approach the initiative with the curiosity and learning mindset of a child, as discussed in the earlier blog.

1. **Leveraging Case Studies and Success Stories:**
   Just as a child learns from examples and models in their environment, Silo AI can showcase real-world examples of how its AI services have transformed businesses. Develop detailed case studies and success stories highlighting how clients have successfully implemented AI solutions to solve specific problems. This not only provides tangible evidence of the value of Silo AI's services but also serves as a source of inspiration for potential customers.
2. **Educational Content and Resources:**
   Silo AI can take inspiration from the way children seek knowledge and learn through exploration. Create and distribute educational content that explains complex AI concepts in simple, easy-to-understand terms. This could include blog posts, infographics, video tutorials, and webinars. By providing valuable resources, Silo AI positions itself as a thought leader in the AI industry, attracting businesses seeking expert guidance on AI implementation.
3. **Interactive AI Tools and Demos:**
   Children learn through experimentation and play. Similarly, Silo AI can offer interacti tools and demos that allow potential customers to experience the capabilities of AI

Regenerate

Send a message

2.6 Give ChatGPT a Role

Experts respond better than ordinary people. The same goes for ChatGPT, if you give it a role: "You are an [X field] expert." Defining a role and area of expertise helps set the tone and expectations for the content produced by artificial intelligence. If you want to refine your prompt, you can define a more precise context.

For example, if you want to make a marketing plan for your company, you can refine the instructions with the commands shown in Figure 2.22.

Note that the latest feature of ChatGPT, so-called custom instructions, gives ChatGPT a role in every conversation you have according to your needs. This way, you only need to define the role once.

Figure 2.22.

You are marketing expert. Here is [background info] of my company. Create a marketing plan with 3 points.
[background info] =
Tailorframe passport photo booths are located around Finland in most Gigantti stores and in selected K-Supermarkets and K-Markets. See the full list of 58 booth locations below on this page.

Tailorframe has been active in the photography industry since it was founded in 2013. Nowadays Tailorframe focuses on developing automated passport photo booths and offering affordable passport photos.

ABOUT PASSPORT PHOTOS
Passport photo 9,90 €

Passport photos taken on a Tailorframe booth are official passport photos. Follow the instructions on the booth to match the Finnish police criteria for a passport photo. The photo is valid for 6 months on the Police License Administration's server. You can use the same photo for multiple applications during the validity period.

We will review your photo on weekdays (Mon-Fri) within 24h between 10 am - 3 pm. Once we have reviewed the photo, we will send it to the Police server and you can attach the photo to your passport application. See list of approved use below on the page.

You will receive the individual, 7-digit retrieval code once your photo is ready to be used on an application.

Tailorframe reviews your photo on weekdays (Mon-Fri) within 24h between 10 am - 3 pm.

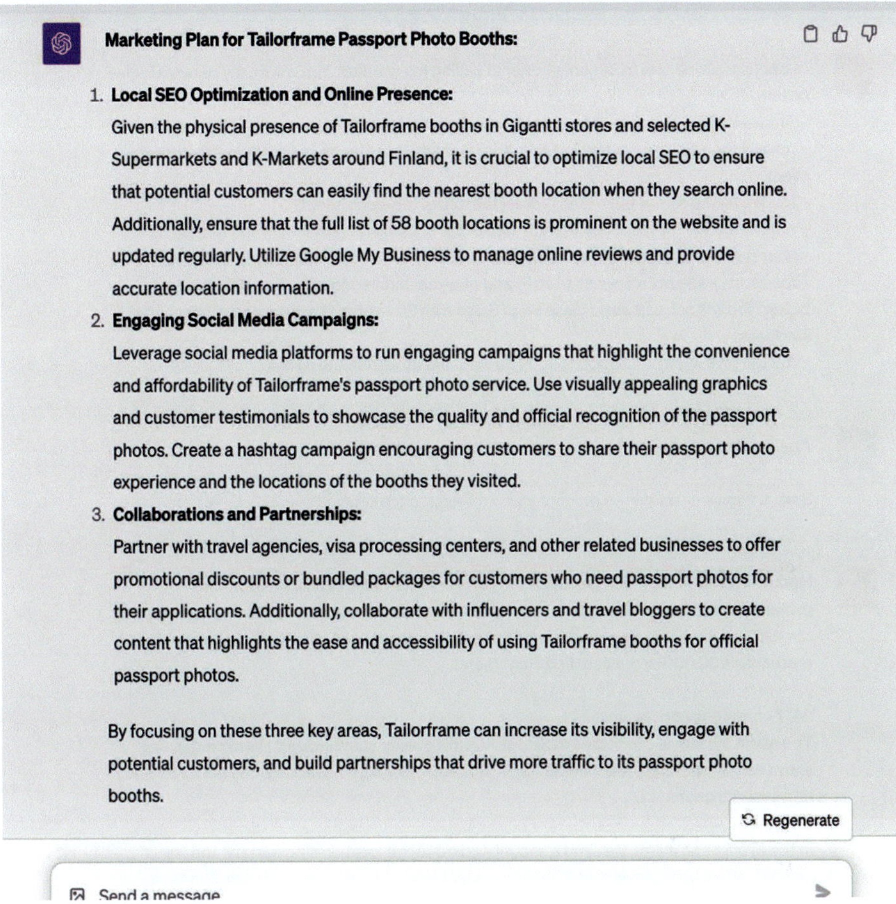

Marketing Plan for Tailorframe Passport Photo Booths:

1. **Local SEO Optimization and Online Presence:**
   Given the physical presence of Tailorframe booths in Gigantti stores and selected K-Supermarkets and K-Markets around Finland, it is crucial to optimize local SEO to ensure that potential customers can easily find the nearest booth location when they search online. Additionally, ensure that the full list of 58 booth locations is prominent on the website and is updated regularly. Utilize Google My Business to manage online reviews and provide accurate location information.
2. **Engaging Social Media Campaigns:**
   Leverage social media platforms to run engaging campaigns that highlight the convenience and affordability of Tailorframe's passport photo service. Use visually appealing graphics and customer testimonials to showcase the quality and official recognition of the passport photos. Create a hashtag campaign encouraging customers to share their passport photo experience and the locations of the booths they visited.
3. **Collaborations and Partnerships:**
   Partner with travel agencies, visa processing centers, and other related businesses to offer promotional discounts or bundled packages for customers who need passport photos for their applications. Additionally, collaborate with influencers and travel bloggers to create content that highlights the ease and accessibility of using Tailorframe booths for official passport photos.

By focusing on these three key areas, Tailorframe can increase its visibility, engage with potential customers, and build partnerships that drive more traffic to its passport photo booths.

Regenerate

Send a message

2.6.1 ChatGPT as a sales coach

ChatGPT can also act as a sales coach, as in the example in Figure 2.23.[21] Note that you can also use an English prompt and ask ChatGPT to respond in Finnish. My son practiced his sales pitch with this exercise and got a summer job in a security company. Many other factors certainly influenced this, but the extra practice was definitely beneficial.

[21] The example is from the Prompt Warrior website: https://www.thepromptwarrior.com/

Figure 2.23.

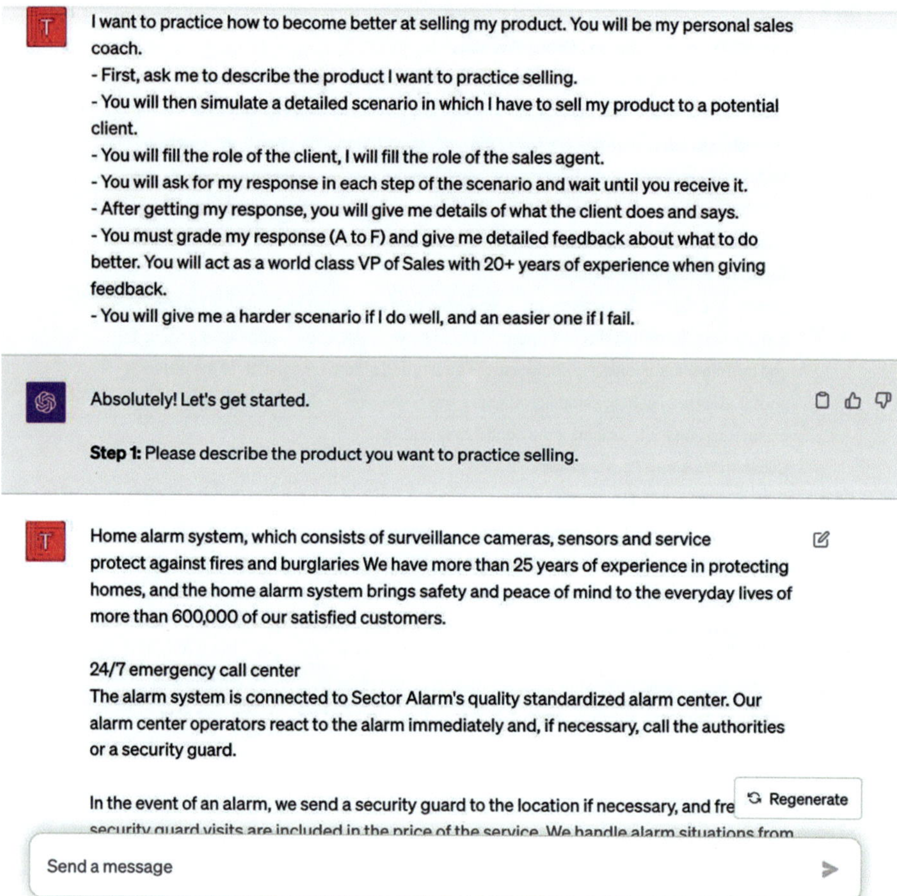

Now we have described the product for sale and the exercise itself can begin. ChatGPT plays the role of a customer. You will notice that ChatGPT gives you a grade for your performance and in addition a detailed description of what went well and what you can still develop (Figure 2.24).

2.6.2 As a ChatGPT strategy director

In the business world, scenarios are often created to support strategic work. Scenarios are descriptions of future possibilities, and they help companies to outline different development paths and their impacts on the company's operations. This way, the company and its management can better prepare for the future and react to upcoming situations.

Figure 2.24.

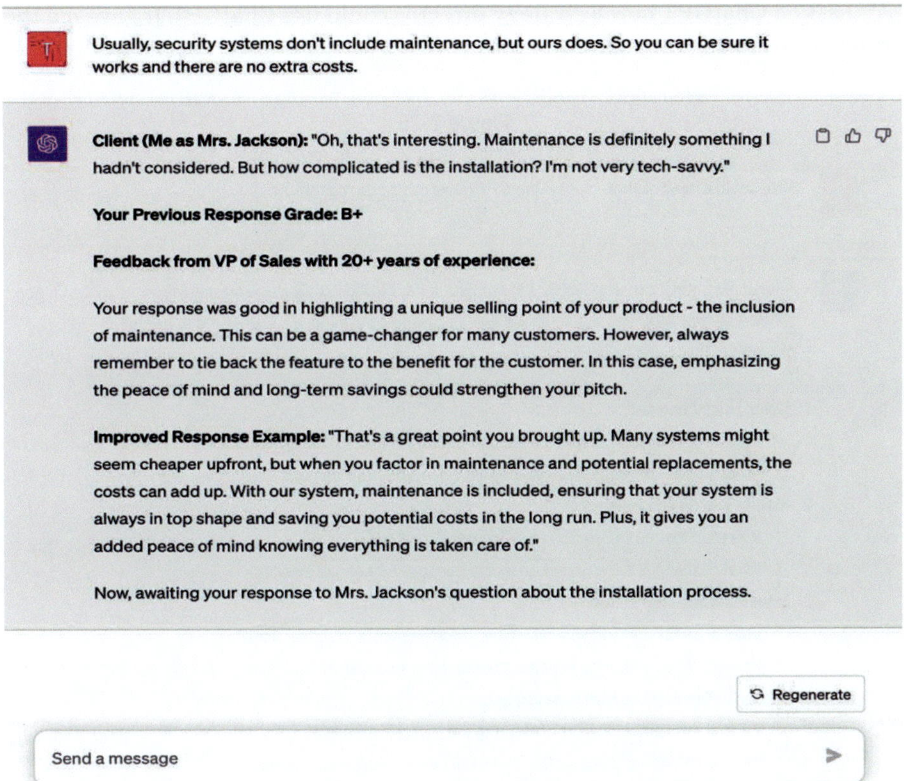

If you want to use ChatGPT to create scenarios, you can share background information with it, such as newspaper articles, blogs, or analyses on the topic you want scenarios for, for example like this: "Create three scenarios [background information]." In the paid version, you can also input web pages. You can say what kind of scenarios you would like to create, what factors should be taken into account[22] and what future trends you want to explore.

[22] Savkin, Aleksey / Medium 27.12.2022: ChatGPT AI for Scenario Planning: Generate Plausible Stories. https://medium.com/@bscdesigner/chatgpt-ai-for-scenario-planning-generate-plausible-stories-34b7730da070

It's important to remember that ChatGPT is a machine learning model and it can't actually think. When creating scenarios based on some background information, it makes mistakes and can even create completely absurd scenarios. Therefore, it's important to critically examine the scenarios and use them only to support your own thinking and ideation.

I asked ChatGPT how a strategy director could use ChatGPT. It gave me the instructions in Figure 2.25.

Figure 2.25.

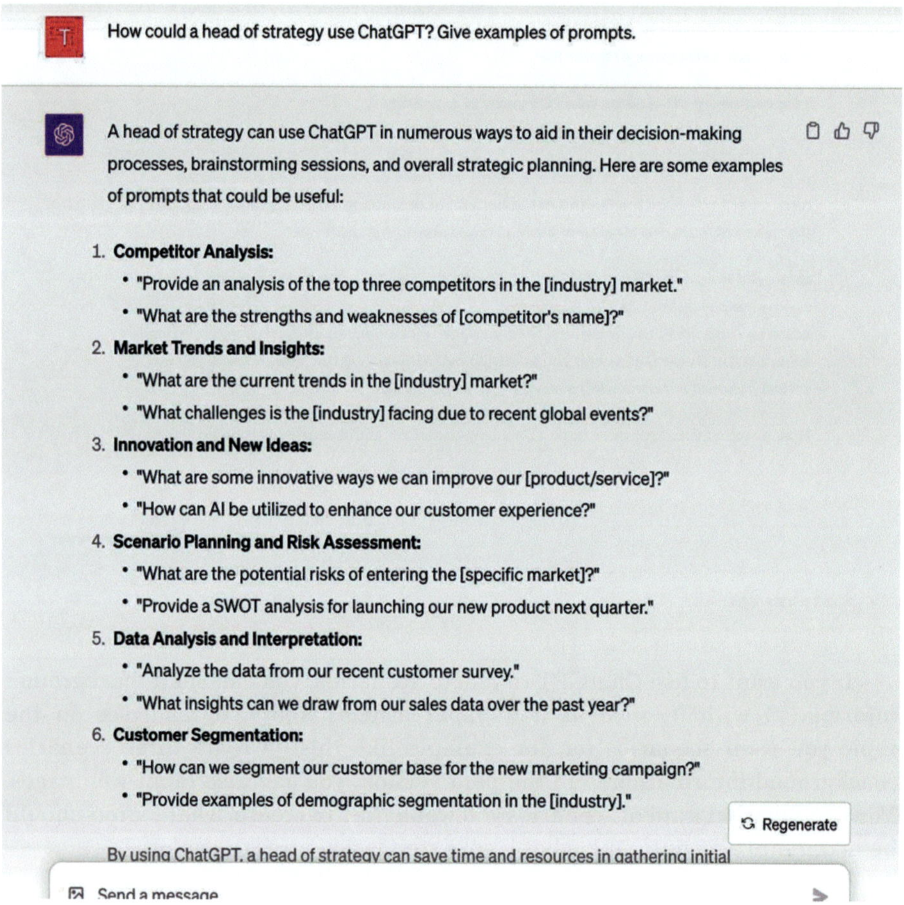

How could a head of strategy use ChatGPT? Give examples of prompts.

A head of strategy can use ChatGPT in numerous ways to aid in their decision-making processes, brainstorming sessions, and overall strategic planning. Here are some examples of prompts that could be useful:

1. **Competitor Analysis:**
   - "Provide an analysis of the top three competitors in the [industry] market."
   - "What are the strengths and weaknesses of [competitor's name]?"
2. **Market Trends and Insights:**
   - "What are the current trends in the [industry] market?"
   - "What challenges is the [industry] facing due to recent global events?"
3. **Innovation and New Ideas:**
   - "What are some innovative ways we can improve our [product/service]?"
   - "How can AI be utilized to enhance our customer experience?"
4. **Scenario Planning and Risk Assessment:**
   - "What are the potential risks of entering the [specific market]?"
   - "Provide a SWOT analysis for launching our new product next quarter."
5. **Data Analysis and Interpretation:**
   - "Analyze the data from our recent customer survey."
   - "What insights can we draw from our sales data over the past year?"
6. **Customer Segmentation:**
   - "How can we segment our customer base for the new marketing campaign?"
   - "Provide examples of demographic segmentation in the [industry]."

Regenerate

By using ChatGPT, a head of strategy can save time and resources in gathering initial

Send a message

2.6.3 Recruitment consultant

ChatGPT can also act as a recruitment consultant. Employees are often evaluated by testing their abilities in various challenges. If you are hiring, you can create assessment tasks for them with the help of ChatGPT. As a job seeker, you can get an idea of how you will be evaluated. The prompt can be used for different roles and abilities, just change the text in brackets according to your current needs (Figure 2.26).

Figure 2.26.

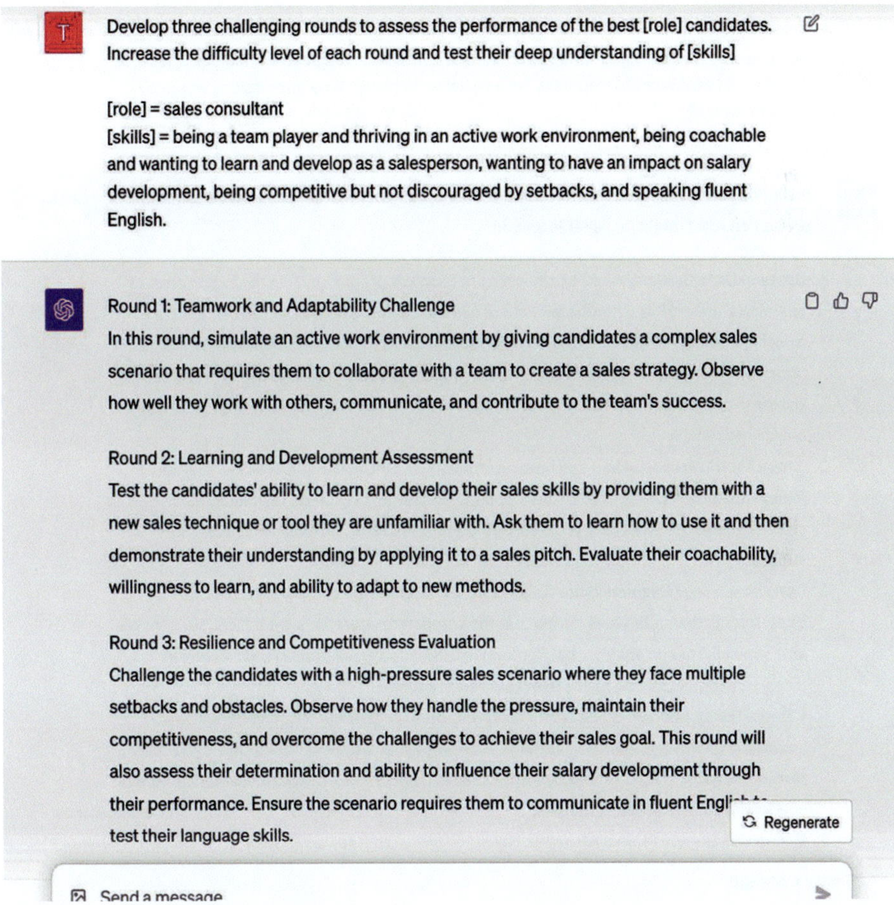

2.6.4 Career counselor

Many young people, and why not adults too, wonder what would be the most suitable profession for them. Artificial intelligence can help with this too, just modify the following prompt to suit you: "I am [describe your degree], who is interested in [what you are interested in]. I am [describe what kind of person you are]" (Figure 2.27).

Figure 2.27.

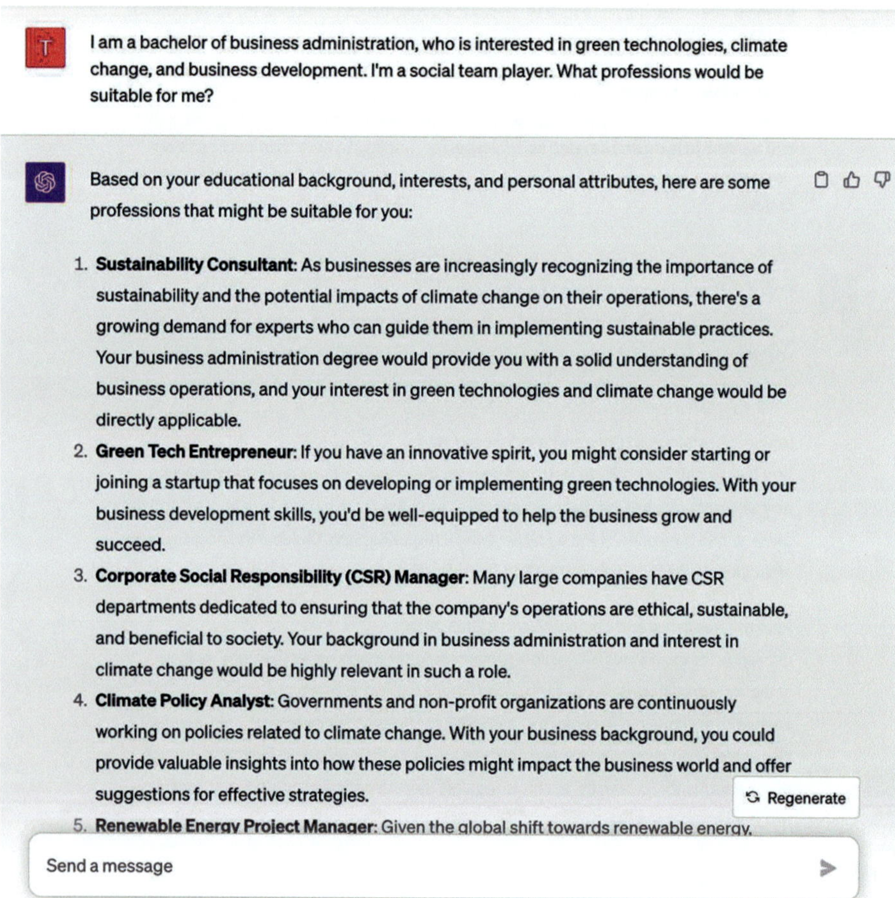
T

I am a bachelor of business administration, who is interested in green technologies, climate change, and business development. I'm a social team player. What professions would be suitable for me?

Based on your educational background, interests, and personal attributes, here are some professions that might be suitable for you:

1. **Sustainability Consultant**: As businesses are increasingly recognizing the importance of sustainability and the potential impacts of climate change on their operations, there's a growing demand for experts who can guide them in implementing sustainable practices. Your business administration degree would provide you with a solid understanding of business operations, and your interest in green technologies and climate change would be directly applicable.
2. **Green Tech Entrepreneur**: If you have an innovative spirit, you might consider starting or joining a startup that focuses on developing or implementing green technologies. With your business development skills, you'd be well-equipped to help the business grow and succeed.
3. **Corporate Social Responsibility (CSR) Manager**: Many large companies have CSR departments dedicated to ensuring that the company's operations are ethical, sustainable, and beneficial to society. Your background in business administration and interest in climate change would be highly relevant in such a role.
4. **Climate Policy Analyst**: Governments and non-profit organizations are continuously working on policies related to climate change. With your business background, you could provide valuable insights into how these policies might impact the business world and offer suggestions for effective strategies.
5. **Renewable Energy Project Manager**: Given the global shift towards renewable energy,

2.6.5 Always ready coach

ChatGPT can also help you improve various documents that you need to prepare. For example, a job application, resumé, speech, or article. It is always ready to offer advice. You just ask, and the coach will give the instructions (Figure 2.28).

Figure 2.28.

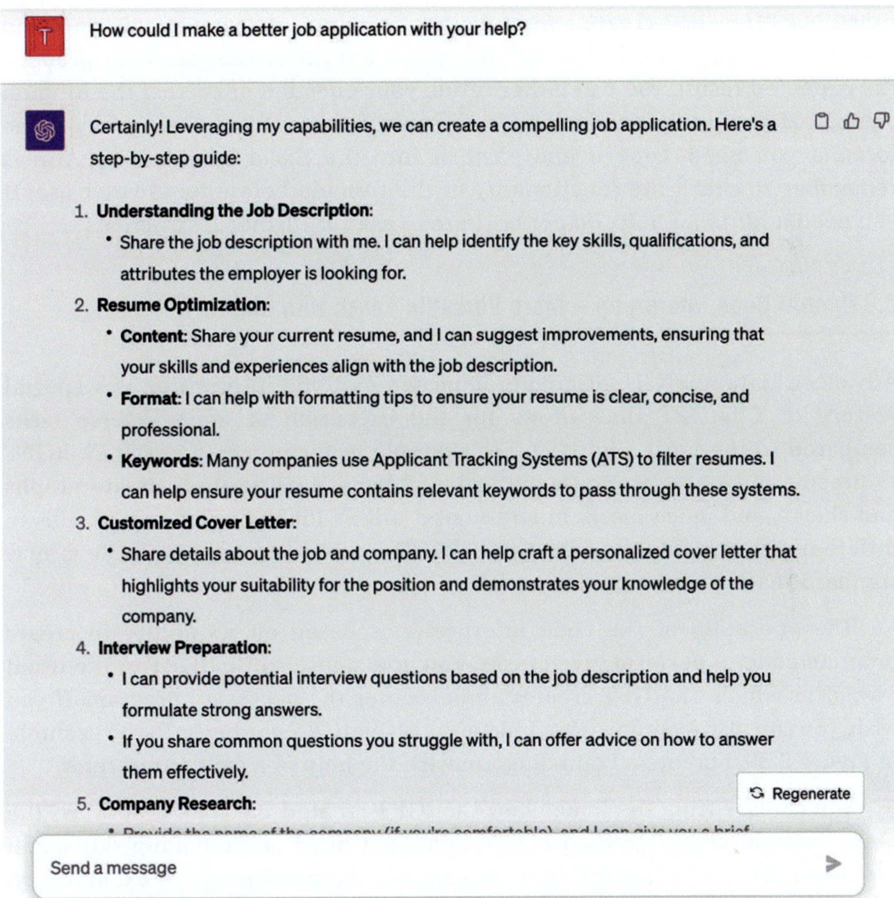

2.7 How do I use ChatGPT with Excel?

Although ChatGPT cannot directly interact with Excel, it can be an extremely useful tool for understanding and creating Excel formulas and functions (Figure 2.29). Use an appropriate prompt, a clear and detailed question, to get the best answer. For example, you can ask "how can I create a formula in Excel that calculates the average for a selected data row". Explain the structure of the table or the data you want to handle, and you will get more detailed instructions.

In addition, you can ask for instructions on creating certain formulas or functions: "How do I create a formula that calculates the sum of column B and column C for each row in Excel?" If you get a formula that does not produce the expected result, you can either refine your question or correct the formula yourself if you understand where the problem comes from. Once you get the formula you need, copy it and paste it into the Excel spreadsheet. Always remember to check the functionality of the formula before its broader use. If you need additional help, do not hesitate to ask for further instructions.

2.8 OpenAI Code Interpreter – More Versatile Tasks with ChatGPT

Advanced data analysis, originally launched as Code Interpreter, is a special feature of ChatGPT that allows for the execution of more diverse tasks compared to the basic ChatGPT. For example, you can use it to ask ChatGPT to process videos, solve mathematical problems, analyze data, create graphs and charts, and much more. In addition, it allows for the uploading of files in different formats directly into ChatGPT. The advanced data analysis is now standard feature of the paid ChatGPT.

The operation of the code interpreter is based on its ability to create program code to perform given tasks. You give a task to ChatGPT in the usual way, after which ChatGPT creates and executes the necessary program. If you wish, you can also examine what kind of code ChatGPT produces. In the example in Figure 2.30, I create a Venn diagram with the help of a code interpreter.

You can see from the image how ChatGPT created the code that draws the Venn diagram visible in Figure 2.31. You don't need programming skills, and the possibilities of ChatGPT have multiplied. You can find more examples on the Medium website, for instance.[23]

[23] SM Raiyyan / Medium 8.7.2023: ChatGPT Code Interpreter: 15 Impressive Real-world Scenarios. https://medium.com/@smraiyyan/chatgpt-code-interpreter-15-impressive-real-world-scenarios-119a9cb036b1

Figure 2.29.

You

Give ten examples of the Excel functions and how to use them, as well as example questions to ChatGPT how to use them

ChatGPT

Certainly! Here are ten Excel functions with examples of how to use them and questions you might ask about them:

1. **SUM**: Adds numbers together.
   - "How do I sum values in Excel from cells A1 to A10?"
2. **AVERAGE**: Calculates the mean of a group of numbers.
   - "What's the formula to get the average of numbers in range B1 to B5?"
3. **VLOOKUP**: Finds things in a table or range by row.
   - "How can I use VLOOKUP to find a specific value in my spreadsheet?"
4. **IF**: Checks whether a condition is met and returns one value if true and another if false.
   - "Can you show me how to use the IF function to display 'Yes' if A1 is greater than 10?"
5. **COUNTIF**: Counts the number of cells that meet a specific condition.
   - "How do I count the number of cells in a range that contain the word 'apple'?"
6. **PMT**: Calculates the payment for a loan based on constant payments and a constant interest rate.
   - "What's the formula to calculate the monthly payment of a loan using PMT?"
7. **CONCATENATE** (or **CONCAT** in newer versions): Joins two or more strings into one string.
   - "How do I join first and last names in two different cells into one cell?"
8. **MAX**: Returns the largest number in a set of values.
   - "What formula should I use to find the maximum value in a list of numbers?"
9. **MIN**: Returns the smallest number in a set of values.
   - "How can I determine the smallest number in my Excel data range?"
10. **LEFT**: Extracts a specified number of characters from the left side of a text string.
    - "How do I use the LEFT function to get the first two characters from a cell's text?"

Figure 2.30.

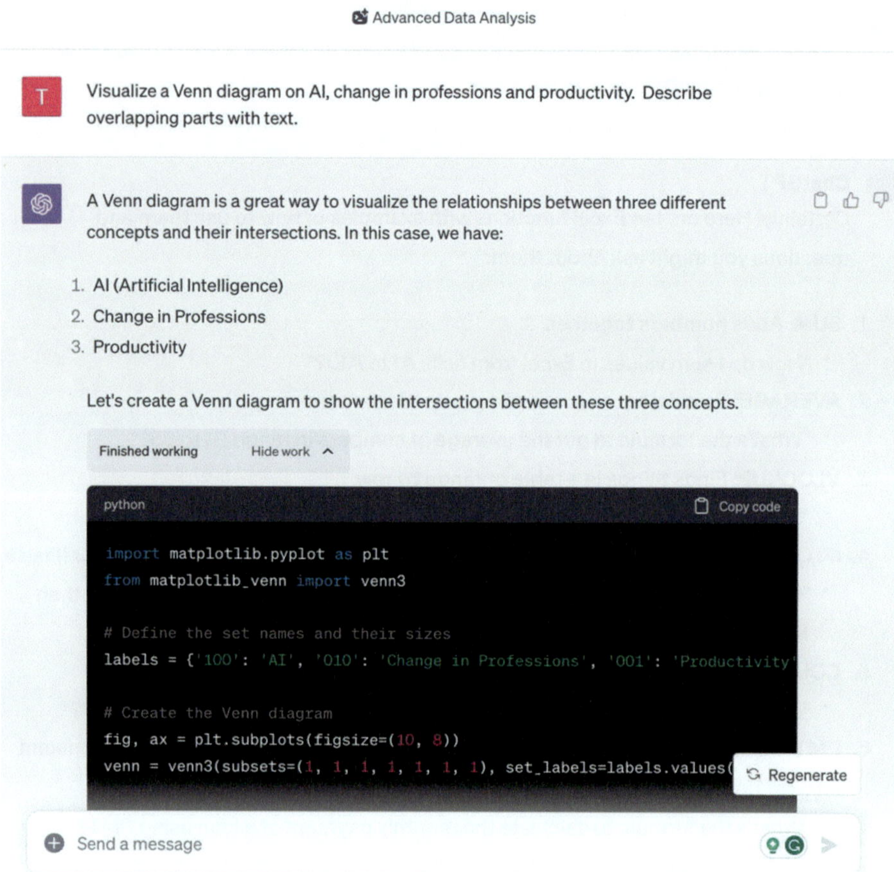

2.9 OpenAI Playground – A platform for Experimenting with AI Models

OpenAI Playground offers users the opportunity to try and test many of OpenAI's artificial intelligence models. The Playground interface is more versatile than ChatGPT's. The Playground has been particularly popular among developers, researchers, and technology enthusiasts, as it provides an easy way to experiment with OpenAI's artificial intelligence models.

Figure 2.31.

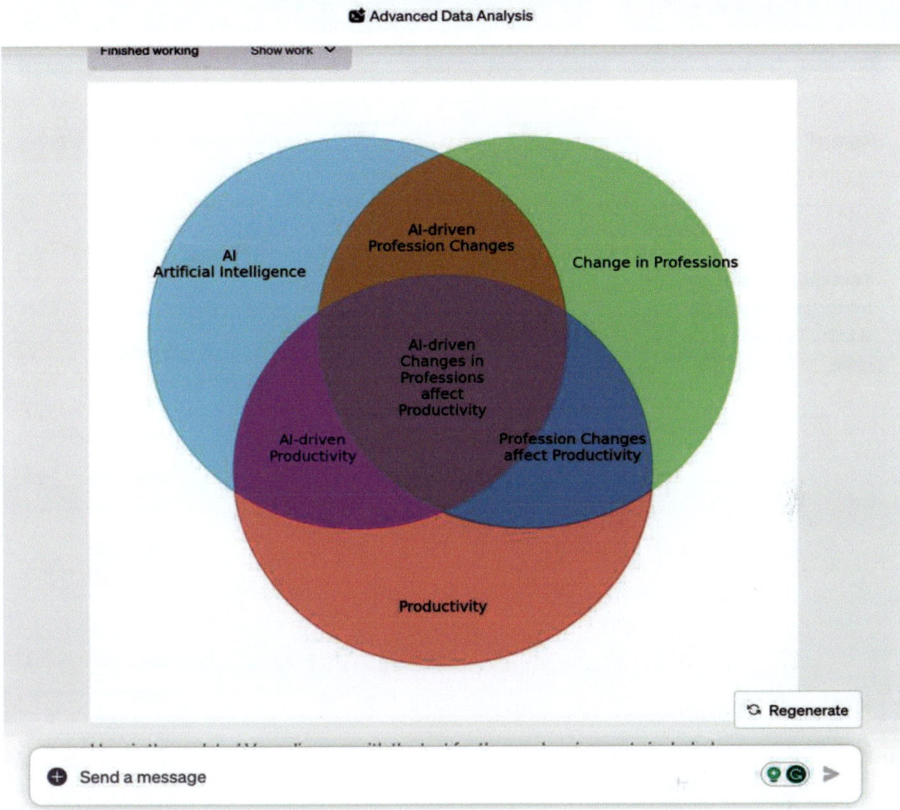

Figure 2.32 shows an example of using OpenAI Playground for language translation. I give it a role in the System section and then a task.

You can choose which model you want to try in the "Model" section and test different inputs and parameters to see how each model reacts. This is also done by professional developers before they integrate OpenAI's models into their own programs. The Playground provides ready-made code for the task, which you can include in your own program. You just press the "view code" button. You can find various examples of Playground usage on the website https://platform.OpenAI.com/examples.

Figure 2.32.

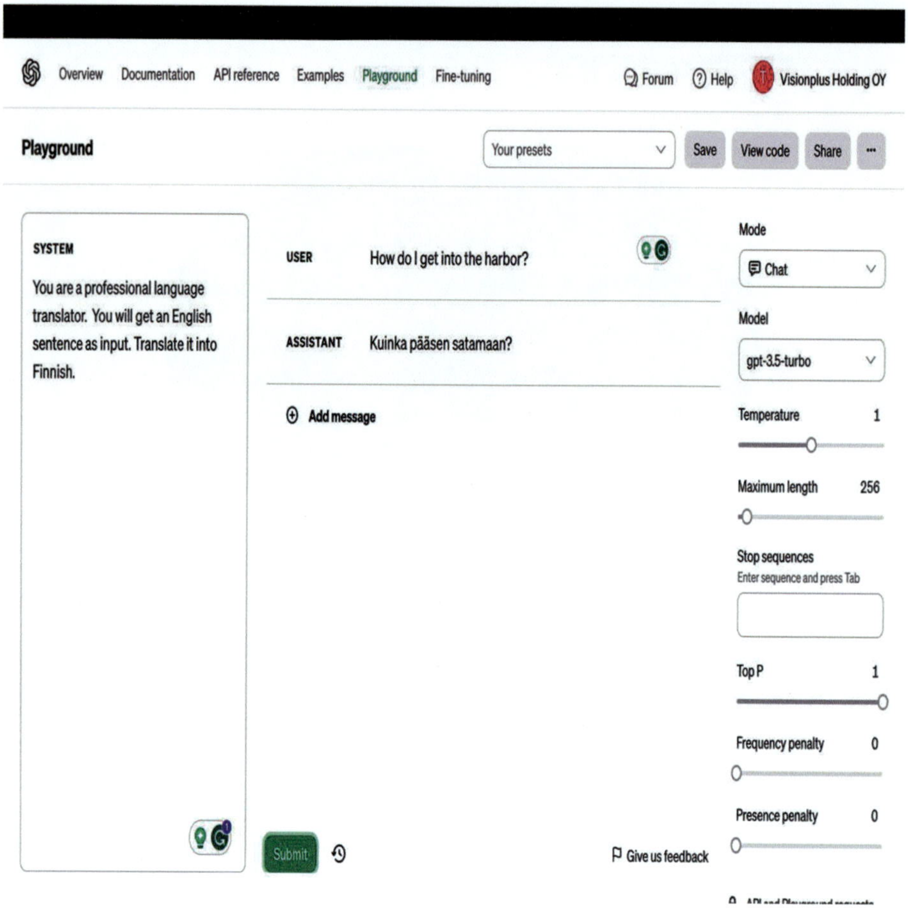

Parameters modify the operation of the models. These are the most important ones:

- Temperature: The temperature parameter regulates the randomness of the model's responses. Higher values (e.g., 0.8) make the result more random, while lower values (e.g., 0.2) make it more deterministic.
- Maximum length: Set the maximum number of characters you want the model to produce in response. This parameter allows you to limit the length of the answer.

- Top P: Determines the diversity of the result. Lower values (e.g. 0.3) make the model choose from a smaller set of likely next characters, leading to more focused responses. Higher values (e.g. 0.9) allow a wider range of possibilities.

2.10 Use of Imaging Software – Midjourney, DALL·E 2 and Stable Diffusion

Next, I will introduce the operation of imaging programs. They produce original, realistic images and art from text descriptions. The most common programs are DALL·E 2, Midjourney, and Stable Diffusion. They can be used for many purposes such as creating art, marketing images, or architecture. For example, you can take a photo of your workspace and use the program to modernize its look with artificial intelligence. If you want to design a new fashion collection, artificial intelligence can assist you in your design process. You can generate logos, characters, buttons, and images for a video game. You can also use artificial intelligence for creating product descriptions.

Next, I will go through the implementation of these applications as well as the structure and operation of prompts. In terms of prompts, I focus on Midjourney, but the basic principles are the same for all applications.

DALL·E 2 seems to be the easiest to use and is best suited for beginners, Midjourney is a bit challenging to implement but offers the best imaging features. It should be noted that users of the free version do not own commercial rights to the images they generate. Stable Diffusion is a free, open-source model, and it can also be installed on your own server if you have the technical skills. You can find a more detailed comparison of these applications on the Product Hunt website.[24] However, remember that all features are constantly evolving, as are the business models of the applications.

You can access DALL·E 2 at https://labs.openai.com/ or its newer version DALL·E 3 in ChatGPT. Just create an account and you can start using it right away (Figure 2.33). The application provides examples of prompts when you click the "Surprise me" button.

You can also upload your own images to the program and edit them by generating variations, removing parts of the image, and modifying the image

[24] Parti, Ayush / Product Hunt/Stories 6.6.2023: Midjourney vs. Stable Diffusion vs. DALL-E: What's best in 2023? https://www.producthunt.com/stories/midjourney-vs-stable-diffusion-vs-dall-e

with prompts. Upload a picture of a room in your house, for example, and modify it by adding furniture, paintings, or other decorative elements.[25]

Figure 2.33.

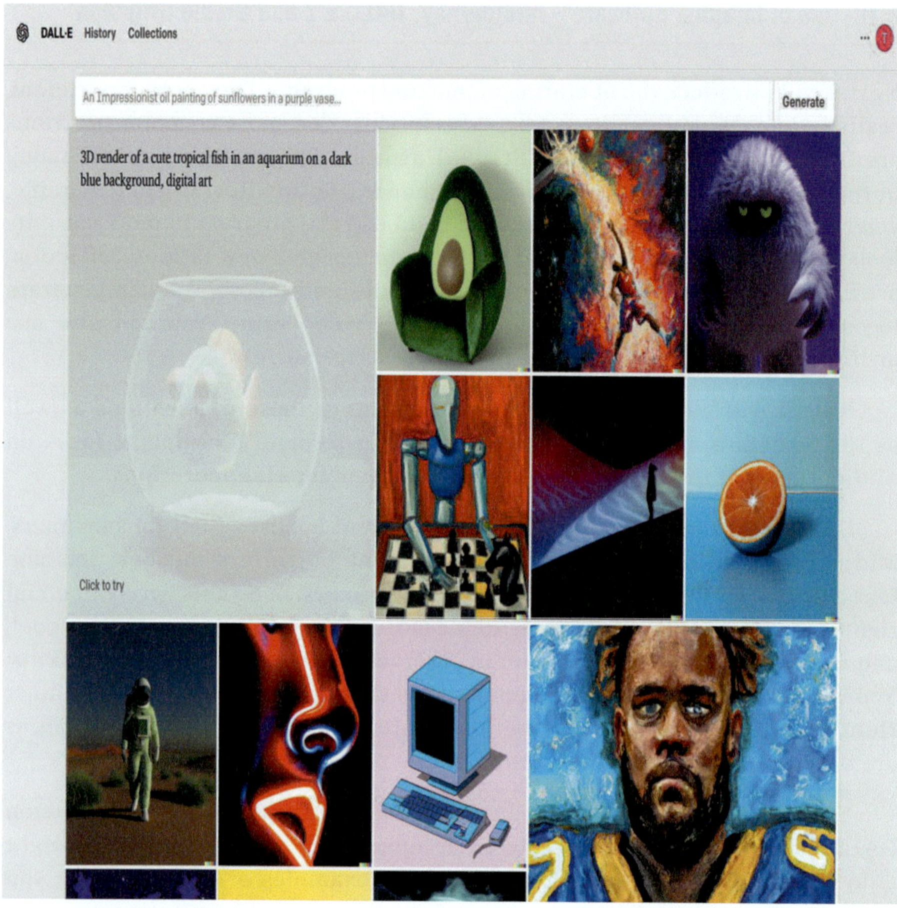

Stable Diffusion is an open source artificial intelligence model, and its web application can be found at https://stablediffusionweb.com/#demo (Figure 2.34).

[25] The Cleverest 31.8.2022: Using DALL-E 2 for interior design: https://thecleverest.com/using-dall-e-2-for-interior-design/

Figure 2.34.

You can also use Stable Diffusion through the Dream Studio application (https://beta.dreamstudio.ai/), where you get slightly more versatile features at your disposal. Just write the prompt and press the Dream button (Figure 2.35).

Stable Diffusion can also be downloaded to your own computer, but it requires some technical skills.

Midjourney is a paid imaging program built on top of the communication platform Discord. Midjourney's developers worked around the world and used Discord for communication. At the same time, they realized that community was the core concept of Midjourney, and built the entire application around it.

I found the implementation of Midjourney a bit tricky, but don't worry, here are the instructions. First, create a Discord account at www.discord.com. The application asks you to create a username, confirm that you are human, and ask for your date of birth. Then you can create a server for yourself, for which you need to provide an email address and come up with a password. Discord will still ask you to verify your email address. After this, type http://discord.gg/midjourney into the browser's address field. This way you can join the Midjourney server, which appears on the left side of the Discord application. After that, you choose a subscription, which is therefore chargeable. You can find more detailed instructions on the Midjourney Quick Start site.[26]

Then you can start using it. Join any channel marked with #newbie or #general. They may initially seem chaotic, as you see the prompts and outputs of all channel users. However, don't be confused, as browsing the channels is a great way to learn what works and what doesn't. Then it's time to try your first command. The most common Midjourney command is "imagine" (written: /imagine), after which you write your prompt, i.e., a description of the kind of

[26] https://docs.midjourney.com/docs/quick-start

image you want to create (Figure 2.36). You can also ask for instructions with the /ask command, such as how to use the source image in Midjourney.

Figure 2.35.

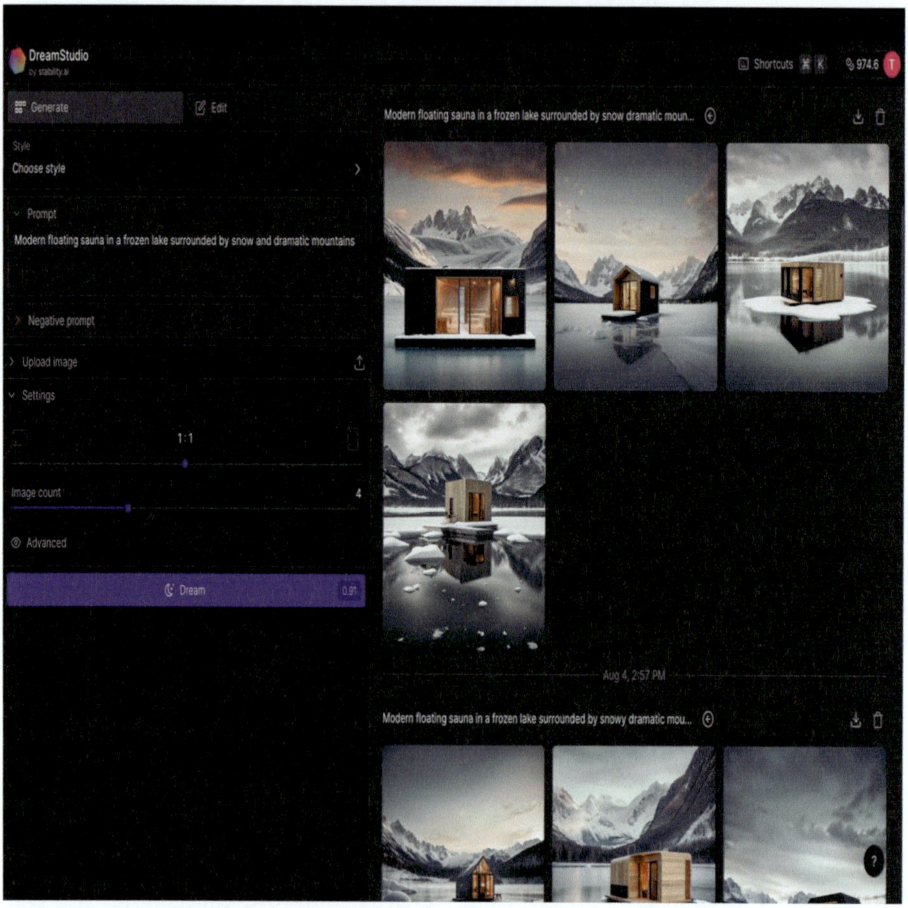

Figure 2.36.

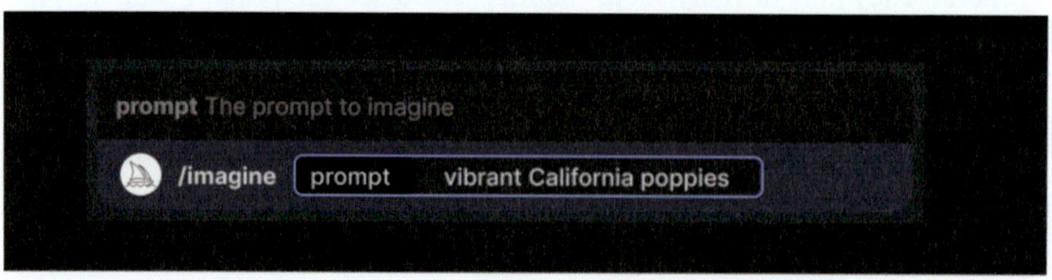

2.11 Prompting in Imaging Applications

Next, I will go through how prompts are used in imaging applications. I will use the Midjourney application as an example, but basic prompts are equally available in DALL·E 2 or Stable Diffusion. For now, the prompt language of these applications is English, in the future there will certainly be more language options.

Just like with ChatGPT, writing prompts for Midjourney and other imaging services is a skill that requires practice. The more descriptive and detailed you are, the more unique and vivid your image will become. Word choice is also important. More figurative synonyms work best. Instead of using the word "big", try "gigantic", "enormous" or "immense". The fewer words you use, the more significant each one becomes. Word order also matters – it's best to place the most important wish at the beginning and the least important thing last. AI models are complex and based on probabilities, so even a small change in the text can dramatically affect the image you create. You learn by experimenting.

It's better to describe what you want, rather than what you don't want. If you ask for a picture of a party without a cake, your picture will likely include a cake. If you want to ensure that a certain object is not in the final picture, you can use the --no parameter. The parameter is always marked with two hyphens without spaces.

Plural words leave a lot to chance, so it's better to use exact numbers instead. "Three cats" is more precise than "cats". It's better to state the number in letters rather than in digits. If you use a number, it can end up in the picture. Collective nouns also work better: so write "flock of birds" instead of "birds".

Think again about what you want as a result. Here are some keywords you can use when creating Midjourney prompts:

- Media: photo, painting, illustration, sculpture, doodle, tapestry.
- Environment: indoors, outdoors, on the moon, in Narnia, underwater, the Emerald City, tundra, salt flat, jungle, desert, mountain, cloud forest.
- Lighting: soft, ambient, overcast, neon, studio lights.
- Colors: vibrant, muted, bright, monochromatic, colorful, black and white, pastel. Colors can be very diverse: millennial pink, acid green, desaturated, Canary yellow, peach, two toned, pastel, mauve, ebony, neutral, day glo, green tinted.
- Mood and emotion: Sedate, calm, raucous, energetic, determined, happy, sleepy, angry, shy, embarrassed.

- Composition: portrait, headshot, closeup, birds-eye view.
- Drawing style: life drawing, continuous line, loose gestural, blind contour, value study, charcoal sketch.
- Time Travel: write a decade or century: 1700s, 1920s 1990s.
- Style: Salvador Dali style, Rembrandt style, Pixar style, aquarelle, oil paint.
- Major cities: NYC, London, Paris.
- Angle: portrait, wide-angle, low-angle.

At the end of the prompt, you can give parameters that change the aspect ratio, the weight of the text and image, and other factors affecting the image. Here are some of the most common parameters and their values. You can find more on the Midjourney website.[27] Remember, you always write two hyphens in front of the parameter form and only then the parameter value or text.

- --ar <number><number> Aspect ratio changes the aspect ratio, the default value is 1:1, other options are --ar 4:3 and for example --ar 16:9.
- --chaos <number 0–100> Make variations and let the model surprise you. A larger number produces unusual and surprising variations.
- --iw <0–2> If you provide an image in the prompt, the image weight value determines the weight of the image in relation to the text.
- --no -parameter removes the given object from the image: for example, "--no plants" would remove the plants from the image.

As imaging applications become increasingly skilled in natural language understanding, the current need for specific words, adjustments, or tricks decreases. Now, prompting is largely about experimenting with what works and what doesn't. And as models update, it may be that previous tricks no longer work.

Since I particularly like modern log cabins, Figure 2.37 shows an example related to them. As a result of the prompt, we get a stunning, bright modern log house bedroom with a view of the mountains.

Prompt: A bedroom of a beautiful bright modern log house cabin with large windows opening onto snowy mountains. The room is well illuminated.

I posted a picture on Instagram with the text "The latest plan for the Lapland cabin". My mother-in-law quickly replied: "Cool your jets." However, with the help of artificial intelligence, one can always continue dreaming – and planning.

[27] https://docs.midjourney.com/docs/parameter-list

Figure 2.37.

In addition to text, you can also use images as a prompt or as part of one. The next image's broader prompt (Figure 2.38) includes the /imagine command, a model image web address, a text prompt about what is wanted to be created this time, and parameters to modify the result.

An easy way to bring an image into Midjourney is to copy and paste it onto the command line with the good old copy-paste function. Remember to press enter at the end so that Midjourney recognizes the image. After that, you can pick up the URL or web address of the image.

Figure 2.38.

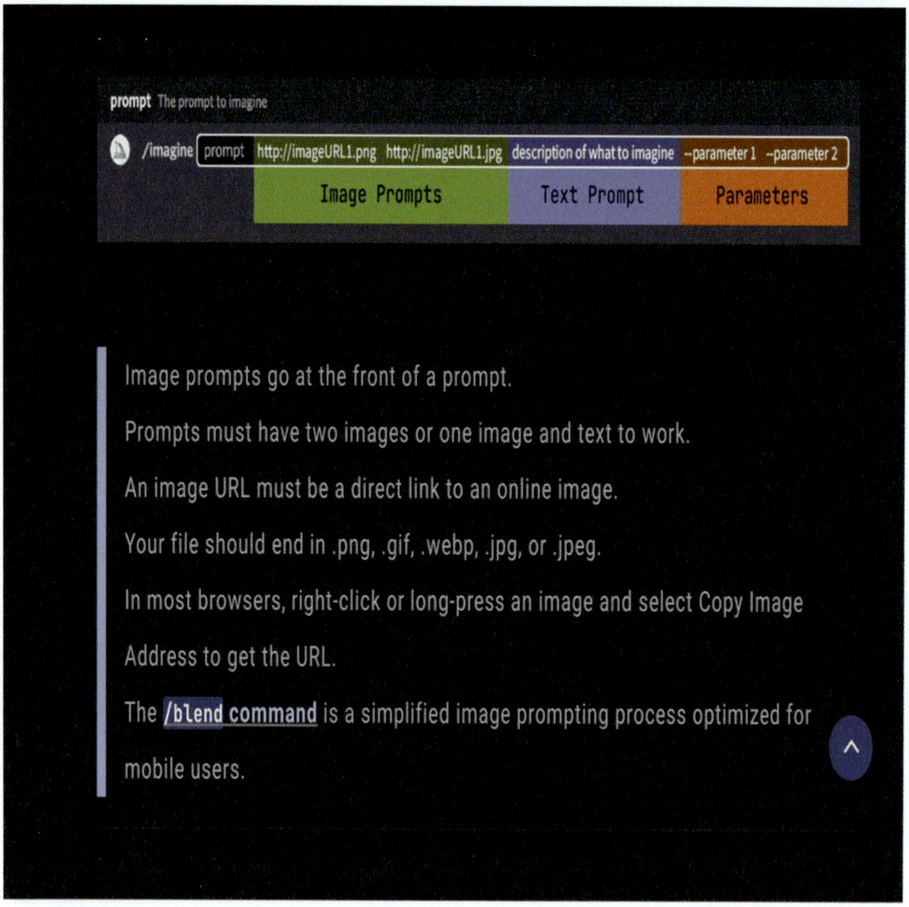

If you want to process multiple images at once, use the /blend command. It allows for quick uploading of 2–5 images, after which the program examines the concepts and aesthetics of each image and combines them into a new, original image. Note that the /blend command does not work with text prompts. If you want to use text and image prompts together, use image prompts and text with the /imagine command.

2.12 Examples of Use

Midjourney and other imaging services can be used for a wide variety of purposes – the only limit is your imagination. Images created in image services can be used on social media, as a basis for fashion design, game design, advertising and marketing, book illustrations, movie design and concept art, product and event design, as well as architecture and interior design.

Here are a few examples from my own experiments.

- Prompt: "Retro Instagram photo of a woman holding ice cream in a summer park."
    - A retro-inspired Instagram photo of a woman in a summer park with an ice cream in her hand.
- Prompt: "Futuristic women's outfit made of shiny metallic fabrics with neon lights."
    - A futuristic women's outfit made of shiny metallic fabrics and neon lights.
- Prompt: "Medieval castle surrounded by a dragon and an enchanted forest."
    - A medieval castle surrounded by a dragon and an enchanted forest.
- Prompt: "Vintage advertisement poster of a 1950s car on a sunny beach."
    - Vintage poster of a 1950s car on a sunny beach.
- Prompt: "Mystical forest with talking animals and magical mushrooms."
    - A mysterious forest with talking animals and magic mushrooms.
- Prompt: "Space station orbiting an alien planet with three suns."
    - A space station orbiting a foreign planet under three suns.
- Prompt: "Eco-friendly plastic-free water bottle made of bamboo and glass."
    - An ecological plastic-free water bottle made of bamboo and glass.
- Prompt: "Outdoor wedding reception with white tents, floral arrangements, and candles."
    - Outdoor weddings, white tent canopies, floral arrangements, and candles.
- Prompt: "Three-story urban townhouse with a rooftop garden and minimalist design."
    - A three-story townhouse with a rooftop garden and minimalist design.
- Prompt: "Rustic countryside villa with stone walls, wooden beams, and a large fireplace."
    - A rustic villa with stone walls, wooden beams, and a large fireplace.
- Prompt: "Scandinavian bedroom design with neutral colours, wooden furniture, and cozy textiles."
    - Scandinavian bedroom with neutral colors, wooden furniture, and cozy textiles.
- Prompt: "Bohemian living room with colourful rugs, hanging plants, and vintage furniture."
    - A boho-style living room with colorful rugs, hanging plants, and vintage furniture.

- Prompt: "Futuristic office building with glass facades, green terraces, and smart technology integration."
  - A futuristic office building with glass facades, green terraces, and integrated smart technology.
- Prompt: "Modern glass house in a mountain landscape with a waterfall."
  - A modern glass house in a mountain landscape near a waterfall.
- Prompt: "Scandinavian living room with a blush velvet sofa and gold accents."
  - Scandinavian living room with a pink velvet sofa and golden decorations.

You can also ask ChatGPT to help with formatting prompts (Figure 2.39).

Figure 2.39.

Haluan suunnitella modernin vuoristomökin Alvar Aallon tyyliin. Mökki olisi vuorilla vesiputouksen lähellä. Luo minulle englanninkielinen Midjourney-prompti ja anna sen loppuun parametrit. Pelkkä lopputulos riittää, ei alku- ja lopputekstejä.

"Alvar Aalto-inspired modern mountain cabin nestled in the mountains close to a waterfall, emphasizing natural materials and large windows." --style "modern architecture" --ar 16:9

I input this prompt into Midjourney and the result was as in Figure 2.40. Pretty impressive, I'd say.

2.13 Adobe's Photoshop and Firefly

When designing the book cover, graphic designer Riikka Rajapuro from Alma began actively testing AI tools for image professionals: Adobe's Photoshop Beta and Firefly applications, which have integrated new artificial intelligence features. Among the large corporations, Adobe has indeed been at the forefront of the development of generative artificial intelligence.

You can access these applications by creating an Adobe Creative Cloud account at https://www.adobe.com/fi/creativecloud.html. The prices are quite high, as the programs are aimed at professionals, but students and teachers get a significant discount.

Figure 2.40.

Once you have created an account and installed the Adobe Creative Cloud software, you can download the PhotoShop Beta desktop application and start using it. It can be found in the main menu under Beta apps. Currently, the language of use is English. Note that Adobe will likely later bring the Beta features to the standard Photoshop application as well.

Photoshop Beta is particularly suitable for editing images. Photoshop is based on image layers, or layers. Layers enable even complex edits without interfering with the original image. For example, you can stack images, text, and vector graphics without the images on different layers mixing with each other.

With generative filling, objects can be generated by selecting an area from the image and describing in the text prompt what you want to add or replace. In the example in Figure 2.41, I chose the gable roof of the boat shed from the image and asked to make it a flat roof (flat roof). The program provides several options from which you can choose your favorite.

You can also expand images with the help of generative fill (generative expand). You could, for example, create more lawn in your yard or extend a hedge – whatever you want to try with images.

Adobe Firefly is part of Creative Cloud. It is partly a direct competitor to Midjourney and DALL·E 2, but Adobe's intention is to improve professionals' creative workflows. There is also a Finnish version of Adobe Firefly available, and it is currently free for Creative Cloud license customers.

Figure 2.41.

With Firefly, you can create images with prompts, just like in other imaging applications. I personally find that editing my own images is also very easy with it. The user interface seems to be more consumer-oriented than Photoshop, which has more versatile features but requires more learning to use.

My grandfather built the wooden boat in the picture in 1953 in Luumäki. It is in the original picture in the boat shed. I uploaded the picture to the Firefly app, where I removed the background with one click. Very convenient. There was a bit of vague fluff left on the bottom of the boat, but it was easy to remove. After this, I gave a prompt to create a new image (see Figure 2.42).

The process and the final result are shown in the Figure 2.43: the original, the background removed, and the prompted final result, of which there are two variations. In the latter, the prompt was "a boat on a lake with waves".

Features in the experimental phase of Firefly, such as "sketch to image" and "3D shape to image", open up new possibilities in image editing, as they transform rough sketches or 3D models into realistic images. Adobe Firefly is also integrated with Discord, where the user gets a community and a platform for discussion and support.

Figure 2.42.

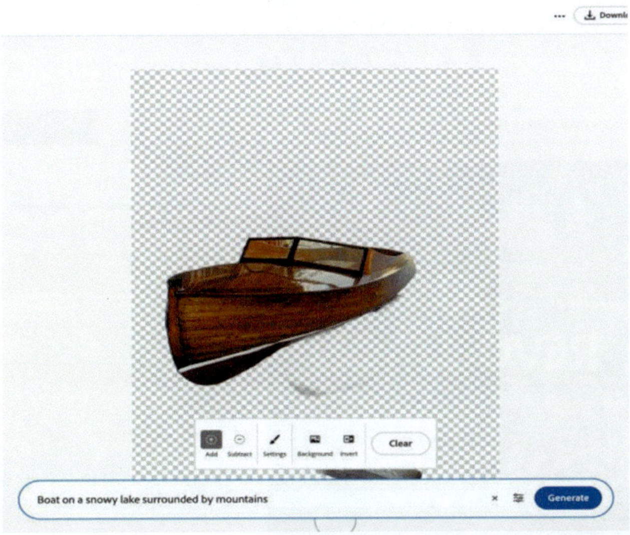

Figure 2.43.

2.14 Text to Images - Ideogram

The problem with image generators has been that they can't produce text within an image. Text in images is found, for example, in guide signs and company logos. Former Google Brain researchers have solved the problem with their Ideogram program. In my opinion, it works even better than other similar image generators, such as Midjourney, DALL·E 2, or Stable Diffusion.

You can find the service at https://ideogram.ai/login. The picture and text I created with it are shown in Figure 2.44.

Figure 2.44.

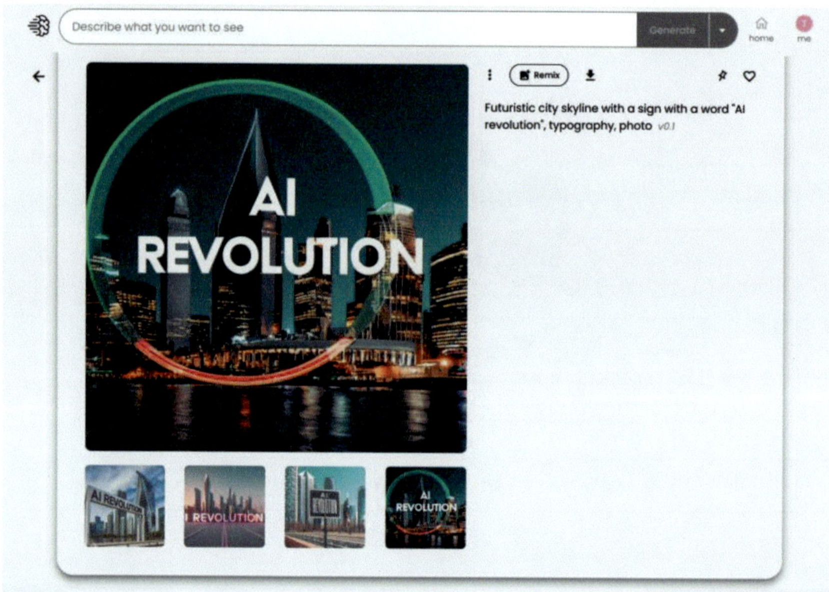

CHAPTER

3

A White-collar Worker's Dream or Nightmare?

The four-day work week has also been a topic of discussion in Finland. The new artificial intelligence enhances creative and knowledge work so much that white-collar workers could take it easier one day a week.

On the other hand, artificial intelligence also destroys professions and job tasks, as technology has always done. Susan, a 25-year-old screenwriter from San Francisco, was replaced by her boss with ChatGPT, and now Susan walks dogs to earn her living.[28] Another professional writer also suffered: his clients started writing their blogs themselves with the help of ChatGPT and terminated their contract with the writer.

It was previously thought that jobs requiring creativity, judgment, and decision-making would be safe from automation. However, the rise of generative artificial intelligence has revolutionized the playing field. Whether you're an experienced expert, visionary leader, coding genius, eager student, inspiring teacher, creative virtuoso, or artistic soul, you cannot escape the impact of artificial intelligence.

Creative work is being automated. Artificial intelligence converses with students like a tutor, and robot lawyers, such as DoNotPay, are changing the work of lawyers. Artificial intelligence creates strategies and scenarios, develops

[28] Verma, Pranshu & De Vynck, Gerrit / The Washington Post 2.6.2023: ChatGPT took their jobs. Now they walk dogs and fix air conditioners. https://www.washingtonpost.com/technology/2023/06/02/ai-taking-jobs/

ideas, and designs products. We must prepare for a future where artificial intelligence not only supports our work, but reshapes entire industries, creating and destroying professions.

It has been predicted that up to 90% of coding will disappear. On the one hand, this means that jobs will be lost, and on the other hand, it means that innovation is cheaper than ever. Many ideas that were previously left unrealized can now be completed because artificial intelligence handles the coding. The amount of work actually explodes.

So, new technology also creates jobs. For example, consulting firm Accenture announced in June 2023 that it is hiring 40,000 new data and AI workers and investing three billion dollars to acquire new capabilities to respond to the disruption of generative artificial intelligence!

In this chapter, I contemplate the impact of artificial intelligence on professions and, through examples, explain which professions change or disappear as technology advances and how the labor market may change in the coming years. And especially: how businesses and employees should respond and react to this upheaval.

3.1 Artificial Intelligence is Transforming Knowledge Work and Creative Professions

The massive base models of the new wave of artificial intelligence are capable of doing things that machines were not believed to be capable of, and therefore their impact on various professions, especially knowledge work, will be significant.

According to the commercial bank Goldman Sachs, generative artificial intelligence can globally enable the automation of up to 300 million full-time jobs. Artificial intelligence is particularly good at office and administrative tasks, legal tasks, architecture and engineering work, and social sciences. Business and financial functions are also susceptible to automation.[29]

According to professors Jukka Luoma and Henri Schildt from Aalto University, artificial intelligence is changing the nature of knowledge work by replacing professions and handling tasks of low abstraction level. This frees up experts' time

[29] Briggs, Joseph & Kodnani, Devesh / Goldman Sachs 26.3.2023: The Potentially Large Effects of Artificial Intelligence on Economic Growth. Global Economics Analyst. https://www.gspublishing.com/content/research/en/reports/2023/03/27/d64e052b-0f6e-45d7-967b-d7be35fabd16.html

to focus on more creative work, which enables a multiple increase in productivity. For example, in software development, a significant part of implementation and testing is transferred to the responsibility of artificial intelligence, which increases the role of software architects in overall management.[30]

Previously, for example, the work of a lawyer was learned by doing routine work under the guidance of senior employees. The trainee drafted, for example, demand letters or settlement proposals under the supervision and guidance of a senior lawyer. When artificial intelligence does the same much faster, those at the beginning of their knowledge work careers can't act as apprentices and accumulate their skills in the same way as before.

Artificial intelligence is changing professions, creating new ones, and affecting how work is done and how it is valued. If artificial intelligence can imitate a human voice, will there still be a need for audiobook readers?

Artificial intelligence companies, such as Banter AI[31] and Forever Voices[32], are capable of imitating any celebrity or private individual, as long as they have enough material for AI training. On Banter AI's website, you can have a conversation with "Donald Trump", for example. So why use a human to create an audiobook?

Now of course someone will argue that a machine cannot fully imitate the unique interpretation and emotional resonance of a human. But can the listener actually distinguish a human from a machine? Meta Company's Voice Box was so good that the company decided to postpone its release because it feared the possibilities of misuse.

The audiobook and reading time service Storytel has already announced that it will bring AI-read audiobooks to the market in partnership with the American company Eleven Labs. According to Storytel, using AI as a reader's voice is 95% cheaper than a human reader. In addition, different language versions can be brought to the service much more easily and quickly than if humans read the books. This way, the service can be expanded to new language areas.[33]

---

[30] Helsingin Sanomat 23.12.2022. Reader's opinion: It's clear that artificial intelligence changes the nature of knowledge work: https://www.hs.fi/mielipide/art-2000009280423.html.

[31] https://banterai.app/

[32] https://forevervoices.com/

[33] Storytel 13.6.2023: Storytel enters strategic partnership with ElevenLabs and announces upcoming launch of new VoiceSwitcher feature. https://investors.storytel.com/en/storytel-enters-strategic-partnership-with-elevenlabs-and-announces-upcoming-launch-of-new-voiceswitcher-feature/

AI-based chatbots can perform a large part of customer service or at least significantly enhance it. Customer service representatives can focus on more complex questions and problems, while chatbots handle routine tasks.

With the help of artificial intelligence, anyone can edit images, choose fonts and color scales, create layout models, and even develop design concepts. This makes graphic design easier and reduces the need for professional skills. The cost of design work decreases when less skills are required from the worker. On the other hand, it gives top designers the opportunity to streamline their work and multiply their productivity.

Already now, algorithms can analyze typography, colors, sketches, and identify inconsistencies and suggest changes. With the help of artificial intelligence, consistent but lively brands can be maintained. Instead of a few fonts being defined for the brand, the fonts can vary according to target group or even customer-specifically. With the help of artificial intelligence, we can move from a static brand to a more interactive brand.[34]

Thanks to artificial intelligence, creating images using the latest AI features of Photoshop is possible for anyone. This makes graphic design increasingly accessible and reduces the need for image processing skills, among other things. This could lead to a decrease in job opportunities for professionals or downward pressure on fees, as the requirements set for workers are lowered. But at the same time, top-class designers can leverage AI more efficiently and produce more, as well as innovate on a whole new level.

For example, graphic designers and designers have used Midjourney to create entirely new types of product concepts.[35] In one experiment, Midjourney was asked to create an image that combined an elephant and a butterfly: the result was a "phantafly". This image was used as inspiration for the Stable Diffusion model, which generated different ideas for product categories, such as chairs.

Architects, on the other hand, can speed up their creative process and gain new types of insights by incorporating imaging services, such as Midjourney,

---

[34] Kauppalehti 9.3.2023. Opinion piece: The brand is an algorithm - the use of AI models changes the way the brand is implemented. https://www.kauppalehti.fi/uutiset/brandi-on-algoritmi-tekoalymallien-kaytto-muuttaa-tapaa-jolla-brandi-toteutetaan/47a26e18-b850-4da4-8bed-22f6e0c19f83

[35] Eapen, Tojin T. , Finkenstadt, Daniel J., Folk, Josh & Venkataswamy, Lokesh / Harvard Business Review July–August 2023: How Generative AI Can Augment Human Creativity. https://hbr.org/2023/07/how-generative-ai-can-augment-human-creativity

into their workflow.[36] By combining, for example, two different style images, Midjourney produces completely new types of sketches and styles.

Will artificial intelligence become a new god? Even priests are not safe from change. In Germany, church services have already been held under the guidance of ChatGPT, with an avatar performing the duties of a priest.[37] Over 300 people attended a 40-minute service with prayers and blessings. A service can be created with a prompt like: "You are a priest, create a service program, include psalms, prayers and a blessing at the end." However, not all those who participated in the service were convinced, as the avatar does not convey emotions. The ability of artificial intelligence to produce human-like characters is developing at a rapid pace, so this may change.

In summary, it can be said that artificial intelligence simplifies complex creative processes by automating them, so that humans can focus on higher-level thinking. But as part of this process, jobs change and jobs also disappear.

3.2 Artificial Intelligence as an Investment Analyst

A venture capital fund was looking for an investment analyst in the spring of 2023, whose role was described as versatile and demanding. Tasks included preparing financing decisions, executing transactions, monitoring and analyzing portfolio companies, and general teamwork, such as producing presentation materials, reports, and analyses. Applicants were expected to have an interest in impact investing and at least 1–2 years of experience. An analytical and goal-oriented attitude were also important qualities. Experience was considered important, but even more important would be the right attitude and the ability to learn quickly.

In the era of artificial intelligence, the last sentence of a job advertisement becomes particularly significant. Artificial intelligence has the ability to prepare financial decisions, summarize potential investment targets, and monitor portfolio companies. In fact, an investment analyst is no longer needed

---

[36] Reinholdt, Eric / 30X40 Design Workshop 30.5.2023: Using AI as design tool in my architecture practice. https://thirtybyforty.com/blog/using-ai-as-design-tool-in-my-architecture-practice

[37] Grieshaber, Kirsten / Associated Press (AP) 10.6.2023: Can a chatbot preach a good sermon? Hundreds attend church service generated by ChatGPT to find out. https://apnews.com/article/germany-church-protestants-chatgpt-ai-sermon-651f21c24cfb47e3122e987a7263d348

for these tasks. Therefore, they must be prepared to learn new things quickly in order to adapt to the changes caused by artificial intelligence.

Venture capitalist and blogger Evan Armstrong has studied the impact of artificial intelligence, specifically OpenAI's GPT-4, on the finance industry as it automates tasks typically assigned to finance professionals.[38] Armstrong believes that AI tools can increase the productivity of finance professionals by up to 50%. According to him, this would reduce the need for entry-level workers in banks, consulting firms, and investment funds.

Armstrong has, among other things, demonstrated that artificial intelligence can analyze a complex financial model in seconds, a task that would take a human significantly longer. He has also shown how quickly artificial intelligence can go through complex financial statement reports.

Artificial intelligence creates value in the financial sector by automating mundane tasks, reducing the need for employees performing executive work. This has been the case before: for example, artificial intelligence has been making loan decisions for a long time. With generative artificial intelligence, even more complex tasks requiring cognitive abilities can be handled by a machine.

On the other hand, as analysis and summaries become easier and cheaper, the demand for services may also increase, as more and more people can afford them. Artificial intelligence thus democratizes services. Previously, the services of a personal financial advisor required considerable capital, but already now there are many kinds of personal digital financial management services available. Generative AI takes this even further, as it can write a personal financial management plan and answer customer questions about it. All automatically, without human assistance.

And if one wants to invest actively themselves, artificial intelligence can create an investment strategy.[39] An investment software called Composer[40] is integrated with the GPT-4 AI model, which it uses to code a functioning investment strategy for anyone. Previously, in the Composer tool, one had to build a flowchart themselves that depicted all the stages of the investment strategy process, which the tool then translated into code. Relatively difficult

---

[38] Armstrong, Evan / Every 18.5.2023: Meet Your New Finance Intern: GPT-4. How AI will reshape what finance professionals do. https://every.to/napkin-math/meet-your-new-finance-intern-gpt-4

[39] Armstrong, Evan / Napkin Math 22.6.2023: AI Can Construct a Trading Strategy Now. https://every.to/napkin-math/ai-can-construct-a-trading-strategy-now

[40] https://www.composer.trade/

and time-consuming. But with the help of artificial intelligence, this also became easier.

An example of an investment strategy prompt could be something like: "Can you build an investment strategy that works both during rising and falling interest rates?" Composer sends this and a supplementary prompt of about 3000 words to the AI, which translates the query into code. It thus automates a previously manual step, the flowchart, using a broad language model (ChatGPT-4) and creates the code to implement the investment strategy. And best of all, it took the company just two weeks to build this feature into their software.

3.3 Democratization of Architecture

I walk along Nuottakallio, a street also known as Finland's Solsidan. A white sugar cube on the right, a similar one on the left. A demolished flat-roofed brick house from the 1970s stood just a moment ago on that beach plot. Architectural styles from different eras are easily recognizable. Even our house is designed by a well-known architect. Could artificial intelligence design houses in the future? Absolutely, and at least the work of architects will change when the new creative abilities of AI tools are implemented.

Existing floor plans are data that can be used to teach artificial intelligence to design new ones. Swedish architectural firm Belatchew has piloted the use of generative artificial intelligence in creating floor plans.[41] The project's goal has been to develop an AI solution for automating the initial phase of the design process. It aims to generate several ideas that can be compared and evaluated based on certain criteria. This initial phase currently takes about 50% of the time of a typical development project when carried out by human architects. An AI-automated solution can reduce the time to a fraction, freeing up architects' time for the next phase of work.

The automation of the initial design phase can be replicated to all stages of the process. For example, in building modeling and simulation. The speed of artificial intelligence in generating different options significantly saves architects' time compared to traditional methods, which in turn shortens project schedules and reduces costs.

[41] About Belatchew architectural firm's AI project on the website of Vinnova, the Swedish government's technology funding organization: https://www.vinnova.se/en/p/creaitor---creative-ai-for-innovative-and-sustainable-architecture/

The speed and capacity of artificial intelligence produce, cost-effectively, various options that a human would not even be able to consider manually. With the help of artificial intelligence, unique architecture can be created that is both energy-efficient and structurally sound.

This could actually lead to an increase in architectural creativity and new innovations.

3.4 Artificial Intelligence Changes the Design Process and Takes Creativity to a New Level

The change affects not only the acceleration of the architect's work but the entire design process.

For example, the world-renowned Zaha Hadid Architects uses Midjourney for creating concepts and draft plans. According to director Patrik Schumacher, working with artificial intelligence is really easy and ideas can be developed together with clients.[42] Artificial intelligence is capable of creating strong and compelling light, shadow, geometry, coherence, gravity, and order in presentation and draft images.

In Schumacher's opinion, using prompts is part of the creative process and thus the outputs of artificial intelligence are creative works in themselves, even though they are created by a machine. In the office he leads, architects choose about 10–15% of the draft images generated by artificial intelligence for the 3D modeling stage.

Prompts often mention the name Zaha Hadid; this is intended to bring the founder's touch, a certain curvature, to the designs. The prompt could be something like: "Zaha Hadid's museum from the air, Baku, high quality." This is an excellent example of human–computer collaboration: AI models learn from architects' previous designs and help them conceptualize new concepts.

The following is an example of a prompt structure that you can use to experiment with different architectural styles:[43] [Architect Name] [Detailed description of subject] [Surrounding environment] [Other styles, aesthetics,

[42] Barker, Nat / Deezen 26.4.2023: ZHA developing "most" projects using AI-generated images says Patrik Schumacher. https://www.dezeen.com/2023/04/26/zaha-hadid-architects-patrik-schumacher-ai-dalle-midjourney/

[43] https://aituts.com/midjourney-architecture/

Designers & Photographers] [Parameters]. The result of my own experiment is an adaptation of a house in the style of Ludwig Mies van der Rohe.

Ludwig Mies van der Rohe, house with glass and mirrors and grid columns, light and modern and transcendent, photographed by Ezra Stoller --ar 16:9 --c 2 --s 90 (Figure 3.1).

Figure 3.1.

More prompts related to architecture and interior design can be found on the website https://promptbase.com/prompt/interior-designs. You can create a room decor or house architecture as easily as writing a Whatsapp message to your friend.

Initially, the pioneers adopt the new technology. Tools like Midjourney require some effort. But gradually, the technology integrates into existing tools, such as Photoshop. This means that consumers can design basic foundation solutions, get 3D visualizations of them, and take their own plan further.

The work that was previously done by professionals becomes possible for the consumer. It is expected that some of the design work will become fully automated, which lowers the threshold for creating architecture. The basic architect's job may even disappear entirely.

On the other hand, this opens up new opportunities for top professionals who can multiply their productivity. Thus, with the help of artificial intelligence, architecture can be both democratized and taken to a whole new level.

3.5 Education and Artificial Intelligence: A Private Tutor in Your Pocket

I practiced Spanish by conversing with a waiter. In a real situation, learning is more fun when you can have a diverse conversation on different topics. This is true – although in this case, it was the new conversational artificial intelligence of the Duolingo language learning app.

With Duolingo's GPT-4-based role-playing game, you can practice different situations in the language of your choice, such as dealing with a cafe or furniture store or planning a vacation. You can also have an immediate feedback discussion with the AI. With the push of a button, you get a simple explanation for the right or wrong answer, additional examples, and even more detailed clarification. In this way, the AI supports an individual learning path. As technology and AI revolutionize teaching and learning, we are far from the traditional classroom.

Another GPT-4-based educational application is Khan Academy, which can be used to support learning. The AI does not provide direct answers, but guides the student step by step to the correct result. It can be used, for example, to support creative writing and history studies. Students create stories together with the virtual assistant Khanmigon, or interview historical figures such as Cleopatra or Jane Austen, making learning interactive and fun.

These AI-based teaching applications work like private tutors, available anytime and anywhere. They recognize mistakes, provide feedback and instructions, and interact with the learner in a way that resembles traditional teaching. AI-based learning platforms are able to adapt to the learner's needs, provide customized instruction, and continually challenge the student, which helps them learn more effectively and quickly.

This does not mean that the role of teachers would disappear entirely. On the contrary, technology can help teachers focus on individual guidance and the special needs of students, while routine tasks, such as practicing grammar, can be outsourced to artificial intelligence.

All of this, however, challenges the current model. Artificial intelligence is changing the world of education in many ways, including how and where learning takes place. With the help of AI, we can make learning a more personal, flexible, and accessible experience for everyone.

3.6 Artificial Intelligence Gets Medical License

In November 2022, a student aiming to become a doctor answered 60% correctly in the United States medical licensing exam. In April 2023, the exam was passed with clear numbers, 90% of the answers were correct. This was not an ordinary student but an improved artificial intelligence model. The April exam was taken by GPT-4.

In another experiment, the Google Med-PaLM AI model analyzed medical claims: 92.9% of its responses were evaluated to be in accordance with scientific consensus.[44] Even though AI diagnoses symptoms better than doctors on average, can we trust it? Not yet, as AI makes mistakes, even if it has been specifically trained with medical data.

However, pioneers want to take advantage of generative artificial intelligence technology, which is only a few months old. For example, in the United States, the UNC Health organization uses artificial intelligence to automatically respond to the most common and time-consuming patient messages.[45] It sounds simple, but previously artificial intelligence was not capable of this. To ensure patient safety, a doctor checks each message before sending. At the same time, the system learns.

The Nuance company, owned by Microsoft, has developed an artificial intelligence system capable of summarizing a doctor's visit based on the conversation between the doctor and the patient. This saves time and effort, and documentation is done thoroughly.

Drug development, disease diagnostics, individual patient treatment plans... the list could go on. It is clear that generative artificial intelligence revolutionizes medicine, its research, and the medical profession. However, there are also challenges in applying artificial intelligence in healthcare, such as the lack of interpretability of results, the need for large amounts of data, lack of transparency, data protection, and ethical issues.[46]

But in the end, the machine is much smarter and more reliable than the average doctor. Do we know how to change healthcare processes and take

---

[44] Raieli, Salvatore / Medium 17.3.2023: Google Med-PaLM: The AI Clinician. https://towardsdatascience.com/google-med-palm-the-ai-clinician-a4482143d60e

[45] Landi, Heather / Fierce Healthcare 25.5.2023: Epic is going all in on generative AI in healthcare. Here's why health systems are eager to test-drive it. https://www.fiercehealthcare.com/health-tech/epic-moves-forward-bring-generative-ai-healthcare-heres-why-handful-health-systems-are

[46] Schuberg, Charlotte / Geekwire 30.3.2023: Microsoft Research head Peter Lee on the applications of GPT-4 in medicine and life sciences. https://www.geekwire.com/2023/microsoft-research-head-peter-lee-on-the-implications-of-gpt-4-for-medicine-and-research/

advantage of this resource? Do we dare to free up resources from routine work for real care?

3.7 Robot Lawyers Invade the Legal Field

AI-assisted lawyers have arrived in courtrooms and law firms. In March 2023, consulting giant PwC and legal tech startup Harvey announced their global partnership. PwC uses AI for tasks such as contract analysis, regulatory compliance monitoring, and due diligence in corporate acquisitions. In the same month, the University of Illinois announced that GPT-4 had passed the Uniform Bar Examination in the United States.[47]

Harvey was the first, but certainly not the last, artificial intelligence-based legal service. It supports human work. PwC indeed states that every result produced by Harvey is checked and approved by PwC's human professionals.[48]

Harvey resembles a cyborg that reads and interprets laws. It can free up human time from time-consuming and routine tasks, such as reviewing documents and the due diligence process. As a result, legal teams can focus on drafting strategic plans and spend more time with clients – maybe even take a break at the coffee machine!

No wonder Harvey is like a popular rock band, with every gig sold out. Major corporations, such as Allen & Overy and PwC, have already bought tickets to the Harvey concert, and over 15,000 law firms are queuing up to get on the next tour. The downside, however, is that many current legal tasks will become redundant, and there will no longer be routine tasks for budding lawyers to hone their skills for increasingly difficult tasks.

But Harvey is just the first wave of the revolution. Around the corner are fully automated services, such as DoNotPay, known as the world's first robot lawyer. It was developed by British Joshua Browder, who started the project to help people avoid parking fines in London and New York. The application quickly expanded to offer a wide range of legal services, examples include airline compensation claims, property tax appeals, and the creation of simple legal documents, such as lease agreements.

---

[47] https://www.iit.edu/news/gpt-4-passes-bar-exam

[48] PwC 15.3.2023: PwC announces strategic alliance with Harvey, positioning PwC's Legal Business Solutions at the forefront of legal generative AI. https://www.pwc.com/gx/en/news-room/press-releases/2023/pwc-announces-strategic-alliance-with-harvey-positioning-pwcs-legal-business-solutions-at-the-forefront-of-legal-generative-ai.html

The legal challenges faced by DoNotPay are evidenced by the fact that the service had to stop litigating court cases and focus on defending consumer rights, such as reducing medical bills, cancelling subscriptions, and disputing credit reports. This was because DoNotPay does not have a law degree. US legislation does not accept the use of such technology, and the founder even faced the threat of imprisonment.

AI-assisted legal services, such as Harvey and DoNotPay, have already brought a breath of change to an industry that has traditionally been routine and static. They make the legal system more accessible, efficient, and dynamic. Trials no longer need to be expensive, and law does not have to be boring.

Artificial intelligence not only changes how the legal system is used, but also how lawyers do their work. Many traditional tasks may disappear, but at the same time new opportunities arise. Legal work may not necessarily decrease, as with the help of artificial intelligence, people have increasingly better opportunities to use lawyers' services – in fact, it may even increase the amount of legal work. Of course, we can ponder whether this is a good or bad thing.

3.8 Artificial Intelligence Creates New Jobs

Although artificial intelligence destroys jobs, it also creates new ones. So, does artificial intelligence help with the current labor shortage? In Finland, in 2023, there was a shortage of professionals in the following fields: practical nurse, registered nurse and public health nurse, specialist in social work, early childhood education teacher, special education teacher, speech therapist, doctor, psychologist, restaurant and institutional worker, waiter, and cleaner.

The labor shortage could at least be alleviated by using artificial intelligence to assist doctors, psychologists, and special education teachers. But artificial intelligence finds it hard to replace personal interaction and emotional intelligence. Therefore, artificial intelligence does not help with the shortage of social workers and mental health counselors. And even though robotics will replace certain physical jobs, the jobs of a waiter, housekeeper, gardener, or nurse are not going anywhere with the advent of creative artificial intelligence.

A new professional category has already emerged, prompt engineers. The task of these people is to create instructions for the use of artificial intelligence. The development of this professional group is analogous to the development of professions that emerged with the internet: before the net, there were no web designers, digital marketing professionals, or search engine optimizers.

Chapter 2 discussed how to create prompts or instructions for artificial intelligence to achieve a desired outcome, whether it be text, an image, or a new piece of music. Developing prompts has become a job in its own right. By instructing artificial intelligence more precisely and intelligently, better results are achieved. Not everyone has time to think about new prompts, so it has already become a new profession for some people.

Companies are also hiring prompt engineers. Their task is to utilize the company's artificial intelligence syste ms more efficiently and to test the systems.[49] Jailbreaking is a term used to refer to tricking, for example, ChatGPT to give answers that are prohibited according to OpenAI's ethical policy guidelines. Clever experimenters have managed to break the protection. Therefore, OpenAI has hired employees to do the same, so it would discover security gaps before they become problems.

3.9 How do I Survive? Options for Individuals and Businesses

The change brought about by the AI revolution does not happen overnight. It took Google five to ten years to revolutionize the advertising market. It took some time before smartphones changed the way we communicate, work, and entertain ourselves. The era of AI is no exception. We have time, but it is important to start adapting to the change now.

At the individual level, it is important to acquire knowledge and skills that are relevant in the era of artificial intelligence. It is necessary to stay up to date with new technologies, acquire new skills, and be open to new learning opportunities.

The AI revolution will immediately impact some professions, such as audiobook readers. Others are protected by legislation and the complexity and diversity of the profession, like doctors. However, these professionals also need to consider how they can best utilize new tools and learn new skills.

Companies must recognize the opportunities and threats brought by artificial intelligence. It is necessary to consider how AI can be utilized in the company's operations and in gaining a competitive advantage. Investments are needed to take advantage of new technology, develop expertise, and train employees. Companies should also note that employees will use these tools

[49] Mok, Aaron / Insider 1.3.2023: 'Prompt engineering' is one of the hottest jobs in generative AI. Here's how it works. https://www.businessinsider.com/prompt-engineering-ai-chatgpt-jobs-explained-2023-3?r=US&IR=T.

anyway. Therefore, they should create guidelines on how to implement them and what is allowed and what is not. This is what Vaisala has done, for example, in the case of AI-assisted coding.

Companies also need to take into account that consumer expectations to understand the AI-assisted operation and management of the services they use have increased. For this reason, the financial company OP has adopted a human-centric AI assessment model and published the first transparency report on the AI it uses.[50] The first transparency report based on the assessment model describes the AI of OP's Financial Balance service.[51]

Politicians and decision-makers need to actively guide the change brought about by artificial intelligence so that it benefits society at large and not just a select few. We need sensible regulation, educational reforms, and investments in future labor markets. We also need to reform existing structures, support flexible professional transitions, and consider scenarios where there may not be jobs for everyone.

[50] OP Group / Cision 21.1.2023: OP introduces a human-centered AI assessment model and publishes its first transparency report on its AI. https://news.cision.com/fi/op-ryhma/r/op-ottaa-kayttoon-ihmislahtoisen-tekoalyn-arviointimallin-ja-julkaisee-ensimmaisen-lapinakyvyysrapor,c3696412

[51] OP Group 2022: Economic Balance - Artificial Intelligence Transparency Report. https://www.op.fi/documents/20556/39794367/OP-Ryhm%C3%A4-Tekoalyn-lapinakyvyysraportti-Talouden-tasapaino-2022.pdf/8403ea51-e911-b40d-a1dd-33d25aff2d93

CHAPTER 4

Productivity Leap – Internet to the Power of 2

Productivity growth is the engine of living standard improvement. When worker productivity increases, they can be paid better wages without employers having to significantly raise prices. As productivity increases, incomes rise and consumers can buy more products and services, which improves the standard of living in the long term.

New artificial intelligence tools have already revolutionized working methods and increased productivity. For example, academic research praised the productivity benefits of ChatGPT just a few months after the service was launched. Consulting firms and research institutes have reported significant effects of generative artificial intelligence on productivity. For example, McKinsey has estimated that the productivity impact of generative artificial intelligence could be over 2.6 trillion dollars annually.[52]

And this is just the beginning. Just as the internet once revolutionized productivity, we now face a leap that could be significantly larger. Although there is a lot of hype around ChatGPT and other AI tools, we may underestimate the long-term productivity growth they bring about.

[52] Chui, Michael, Hazan, Eric, Roberts, Roger, Singla, Alex, Smaje, Kate, Sukharevsky, Alex, Yee, Lareina & Zemmel, Rodney / McKinesy Digital14.5.2023: The economic potential of generative AI: The next productivity frontier. https://www.mckinsey.com/capabilities/mckinsey-digital/our-insights/the-economic-potential-of-generative-ai-the-next-productivity-frontier#introduction

Often the adoption of new technology requires significant investments from companies. But a conversational user interface based on generative artificial intelligence's natural language democratizes access to artificial intelligence, and companies and users were able to easily adopt the new technology. Therefore, generative AI models spread very quickly – they gained a hundred million users in just two months. The popularity also led to a large number of new applications. Even very useful ones, features like Adobe Photoshop's Generative Fill, are now appearing in different programs like mushrooms in the rain.

We are striving for a productivity leap that will only accelerate as the use of generative artificial intelligence becomes more common. And the best part is, the success of companies effectively utilizing technology directs resources, such as available labor, in the best way for productivity growth. Productivity improves when successful companies thrive and poor ones wither away.

It is particularly important to ensure that businesses and society immediately adopt new artificial intelligence tools. When the economy grows, well-being increases. In the long run, productivity is the most important factor for growth, and the most important single factor for productivity is the use of technology. When the same outcome is achieved with fewer resources than before, productivity increases. The productivity growth achieved with generative artificial intelligence is also economically very significant. According to McKinsey, artificial intelligence can increase annual global gross domestic product by 7%.[53]

4.1 Compound Interest – Exponential Growth of Productivity Leap

The leap in artificial intelligence productivity can be compared to the compound interest phenomenon. When artificial intelligence is used to speed up and simplify simple things, its impact multiplies over time. Like interest that grows with the original capital, artificial intelligence has the ability to increase productivity and efficiency over time.

Applying artificial intelligence to tasks that seem simple is constantly generating small improvements. Although summarizing an article sounds

[53] Chui, Michael, Hazan, Eric, Roberts, Roger, Singla, Alex, Smaje, Kate, Sukharevsky, Alex, Yee, Lareina & Zemmel, Rodney / McKinesy Digital14.5.2023: The economic potential of generative AI: The next productivity frontier. https://www.mckinsey.com/capabilities/mckinsey-digital/our-insights/the-economic-potential-of-generative-ai-the-next-productivity-frontier#introduction

trivial, it was previously difficult for a machine; this capability was not easily available. And as artificial intelligence is used more and more and its areas of use expand, the impact grows exponentially. Like the compound interest phenomenon, where interest is added on top of the original capital and then new interest accrues on the interest, the effects of artificial intelligence also strengthen over time.

The more artificial intelligence is used in various industries, the more it learns and develops. Systems using artificial intelligence constantly gain new information and experience, which improves their performance and capabilities. This in turn creates new opportunities and innovations that would not have been invented without artificial intelligence.

There are already countless examples of the benefits of artificial intelligence in simple tasks. AI tools make translation work faster and easier. They provide high-quality translations in different languages, saving time and effort. Automatic captioning of images increases the comprehensibility of visual content and improves accessibility for different user groups. Marketing professionals can leverage AI in creating visual content and thus increase the attractiveness of their content. Authors, bloggers, and marketers can speed up their content production process with AI and improve the quality of their content.

Artificial intelligence tools, such as ChatGPT, can also help refine ideas. They offer creative suggestions, different perspectives, and help prioritize the best ideas. Pioneers have already started and are reaping the productivity benefits of the new technology.

4.2 Productivity Benefits of Artificial Intelligence across Different Industries

McKinsey has studied the impact of artificial intelligence on productivity in customer service, sales and marketing, software design, and research and development.[54]

Chatbots and virtual assistants can provide fast and personalized service to customers around the clock. For example, in the tourism and hotel industry, artificial intelligence assists customers in making reservations and provides recommendations on travel destinations.

[54] McKinsey Digital 14.6.2023: The economic potential of generative AI: The next productivity frontier. https://www.mckinsey.com/capabilities/mckinsey-digital/our-insights/the-economic-potential-of-generative-AI-the-next-productivity-frontier#/

Artificial intelligence can be used to produce creative content for marketing and sales. Algorithms analyze large amounts of data and provide accurate customer profiles and personalized recommendations. This way, marketing communication can be better targeted and sales processes can be made more efficient in various industries.

Artificial intelligence helps software developers to speed up and streamline their work. Generative AI models, like Codex, produce code and create complex software. This saves time and allows for the development of innovative applications more quickly.

Artificial intelligence opens up new possibilities in research and development. Algorithms can analyze vast amounts of data and identify critical trends as well as discover new development opportunities. For example, in medicine, artificial intelligence helps in identifying diseases and developing new treatment methods.

Even these examples show that artificial intelligence has enormous potential to improve productivity in various industries. Let's analyze these a bit more closely.

4.3 Chatbots and Virtual Assistants

On a technical level, customer support fits well with current generative AI tools.

From the perspective of artificial intelligence, customer conversations can be thought of as a series of problem-solving, where the optimal order of actions is sought. For example, when a customer says "I can't log in", the AI has to identify what underlying problems are most likely preventing the customer from logging in, and consider what solutions usually help solve these problems – "Could you check if Caps Lock is off?"

At the same time, artificial intelligence must take into account the customer's emotional reactions and use language that increases the likelihood that the customer will respond positively – "That wasn't stupid at all, I always forget to check that too!" Since customer service conversations are widely recorded and digitized, pre-trained large language models (LLM models) can be fine-tuned for customer service with plenty of examples from both successful and unsuccessful conversations.

The American economic research institute National Bureau of Economic Research published a study in April 2023 on the impact of generative artificial intelligence on productivity in AI-assisted customer service.[55] According to it, AI assistance increases employee productivity by 13.%, which is due to shorter conversation time with the customer, a larger number of conversations per hour, and a larger proportion of successfully resolved conversations.

Artificial intelligence assistance particularly benefits less skilled and inexperienced workers, and new employees develop faster on the experience curve with the help of AI. Highly educated workers benefit less, as their tacit knowledge is already part of the AI recommendations content. However, AI improves customers' attitudes towards employees and reduces difficult calls. Employee turnover decreases especially among new employees.

4.4 Website Design

Productivity increases when the same thing is done faster, for example, designing websites with the help of artificial intelligence. The Uizard Autodesigner tool allows the creation of websites with just a few prompt sentences. The AI uses the given information and automatically generates a proposal for the website's appearance. This AI-based tool saves time and streamlines the design work. Uizard Autodesigner makes website design fast, easy, and intuitive: it offers designers the opportunity to create functional user interfaces without deep technical knowledge.

4.5 Software Work Productivity

GitHub Copilot, based on OpenAI's generative AI model Codex, significantly improved programmers' performance. According to the study, programmers who had access to Copilot completed tasks over 50% faster than those who did not use the tool. Less experienced programmers, older programmers, and those who program for several hours a day benefited the most from the tool.[56]

In 2021, over 4.6 million people were working in computer and mathematical fields in the United States. This category includes computer programmers, data

[55] Brynjolfsson, Erik, Li, Danielle & Raymond, Lindsey R. (2023): Generative AI at Work. NBER. https://www.nber.org/papers/w31161

[56] Kalliamvakou, Eirini / GitHub 7.9.2022: Research: quantifying GitHub Copilot's impact on developer productivity and happiness. https://github.blog/2022-09-07-research-quantifying-github-copilots-impact-on-developer-productivity-and-happiness/

analysts, and statisticians. These workers earned a total of $464.8 billion, which is about 2% of the US gross domestic product. If the results of this study were extended to cover the entire population, over 50% growth in productivity would mean significant cost savings in the economy and would have a considerable impact on the growth of the gross domestic product.[57]

Productivity in software work increases significantly when language models are integrated into software development environments. This way, they become part of the programmer's toolkit. At the same time, the threshold for adoption decreases, and thus people without industry training can program. Similarly, even with minimal programming skills, it is possible to perform considerably more complex programming tasks.

The development of generative artificial intelligence will ultimately lead to programming being done with natural language commands. In this case, a person says what they want, and the machine performs the actual programming. And productivity multiplies.

4.6 Productivity of Consulting Services

Generative artificial intelligence revolutionizes consulting services by enabling broader and better advice for companies. AI-based tools can, for example, analyze large amounts of data and provide comprehensive reports that help companies make better decisions and develop their business.

Artificial intelligence can consolidate a large amount of information and provide better advice, which improves business productivity. For example, artificial intelligence can analyze complex market trends and customer data and provide valuable insights and recommendations that help companies optimize their operations and achieve better results.

Artificial intelligence does not replace human consultants, but acts as their partner, complementing their expertise by offering new perspectives. For example, a consultant can use the analyses and forecasts made by artificial intelligence in their work, which strengthens their decision-making ability and helps to provide even better services to customers.

Generative artificial intelligence opens up entirely new possibilities for improving decision-making and enhancing the performance of companies in

[57] Peng, Sida, Kalliamvakou, Eirini, Cihon, Peter & Demirer, Mert 13.2.2023: The Impact of AI on Developer Productivity: Evidence from GitHub Copilot. 2302.06590.pdf (arxiv.org)

the field of consulting services. AI-based tools can simulate different scenarios and evaluate their impact on the company's financial performance, thus helping companies make strategically justified decisions in future planning.

4.7 How is the Productivity Benefit of Artificial Intelligence Achieved? The Role of Society, Workers, and Companies

Generative artificial intelligence enables significant productivity benefits across various industries. Society, companies, and workers play a crucial role in achieving and maximizing these benefits.

The role of society is to create the framework and conditions for the efficient use of artificial intelligence. It is particularly important to promote education and skills development so that people can take advantage of the opportunities offered by artificial intelligence and adapt to a changing work environment.

The time to act is now. Companies must take a leading role in utilizing artificial intelligence and managing its impacts. They can adopt artificial intelligence services or invest in artificial intelligence technologies and develop their own artificial intelligence solutions to improve their productivity.

Companies can innovate their products with the help of artificial intelligence or streamline their operations by integrating artificial intelligence into their processes to smooth or automate routines. At the same time, companies must ensure that employees are provided with adequate training and support to effectively utilize artificial intelligence.

Artificial intelligence does not replace workers, but rather supports and enhances their work. It's important for workers to be open to new technologies and learn to utilize AI tools and assistive features. In addition, workers need to continually learn new skills and update their expertise to match the changes brought about by AI. Training and retraining programs can help workers adapt to the rapidly changing work environment.

CHAPTER

5

AI Changes Society

The Renaissance (14th–15th centuries) was a period of cultural and intellectual transition in visual arts, literature, music, sciences, and philosophy, when the humanistic values of the classical era were once again honored, leading to a tremendous intellectual and cultural flourishing. In many ways, it was about a new way of thinking, new insights, and new abilities of mankind.

The Renaissance era gave us new tools, such as the printing press. The printing press technology was revolutionary, it brought books within the reach of people and increased the spread of knowledge in society. This also led to the birth of newspapers, which enabled the rapid spread of current news to wide audiences.

In the same way, generative artificial intelligence gives us new tools for producing, analyzing, and sharing information and will lead to new innovations that are difficult to predict but will fundamentally shape society.

In this chapter, I examine the societal impacts of artificial intelligence, especially how AI can empower us positively, create equality, and open up new opportunities. Politician and diplomat Henry Kissinger has compared the development of AI to the Enlightenment, to how it changed our way of seeing the world.

The rapid development of generative artificial intelligence brings with it a new kind of complexity and unpredictability to our future. As these systems evolve, they begin to produce results that even exceed the expectations of their creators. This is a situation that is both exciting and challenging at the same time.

I also examine the negative impacts of artificial intelligence, such as the possibilities of misuse of information, criminal use, and even the rise of dictators.

I will go through the societal impacts of generative artificial intelligence as we understand them at the moment, but as the Age of Enlightenment and the Renaissance led to developments that we could not predict, these too are just a scratch on the surface. However, the message to decision-makers is clear: prepare for changes, don't just stand by. Use these tools and guide legislation in a direction that enables things rather than prevents them, of course taking care of safety.

Often, the people who would benefit most from technology resist it the most; they find it strange and cumbersome. Now that artificial intelligence is becoming democratized and is available to everyone through common computer software, precisely those people who would benefit most from it should be able to use it. Society has a responsibility here: it must encourage people to use artificial intelligence and provide opportunities for it, not leave its exploitation as a privilege for the rich and knowledgeable.

Here is something for politicians and officials to ponder: What are you going to do today so that the less skilled can benefit from technology? Do we leave the discovery of opportunities on the shoulders of individuals, or do we as a society seize the opportunity to significantly improve the quality of life?

5.1 Conversational Interface Democratizes Artificial Intelligence

The greatest innovation of ChatGPT has not been in the AI models, but in how those models have been tuned to converse in natural language. Anyone is capable of interacting with the AI and can get it to perform increasingly complex tasks. The conversational interface has spread like wildfire as people around the world have seized the opportunity.

The democratization of artificial intelligence means bringing the power of AI to everyone. It makes AI technologies more accessible, affordable, and user-friendly to a wider audience, regardless of technical education. A conversational interface does just this.

The democratization of artificial intelligence enables more and more people and organizations to utilize AI, for example in decision-making, product development, service production, and research. This leads to new innovations, improved efficiency, and increased social equality, as the benefits of AI are not limited to large technology companies or particularly skilled experts.

While we should be concerned about the risks of artificial intelligence, we should be most concerned about whether we are taking advantage of the opportunity brought by the democratization of artificial intelligence. If we don't, we are actually creating inequality. Therefore, education, accelerated adoption and experimentation of artificial intelligence in both the public and private sectors are needed.

This is a broader issue where we also need to consider how we democratize the data and infrastructure used in AI training. More on this in Chapter 8.

5.2 Artificial Intelligence is Supportive Intelligence

Artificial intelligence is revolutionizing our world in many ways; the most significant impacts are the opportunities it offers to people living with various constraints. AI applications like ChatGPT provide opportunities for communication, information acquisition, and increased independence for people with a disability or learning and understanding challenges.

For example, the Danish startup Be My Eyes has developed an application that allows sighted volunteers to help blind or visually impaired individuals in daily tasks that require vision. The company has now taken a new step in development and offers a "Virtual Volunteer" feature powered by GPT-4, which is capable of producing the same level of context and understanding as a human volunteer.

Traditional image recognition applications only say what is directly in front. An image recognition model implemented with GPT-4-based artificial intelligence can have a conversation with the user, for example, about whether one can trip over an object on the ground. The Be My Eyes app can also identify essential parts from, for example, a website and summarize the search results in the same way as sighted people naturally scan them when reading websites. This way, artificial intelligence helps those who need visual assistance to shop on websites, which are often confusing and contain many elements.

Communication is vital for everyone. AI chatbots can make it significantly easier for individuals who have challenges with hearing or speech. These chatbots enable speech to be converted into text and vice versa, which in turn allows participation in conversations and social interactions.

In addition, ChatGPT can make reading and writing easier for individuals with visual challenges, dyslexia, or fine motor skill challenges. By reading aloud text-based content or providing writing suggestions, ChatGPT helps users create and format emails, essays, and other documents more easily and efficiently. ChatGPT can also be asked to explain an idea or curriculum in a specific way to a person with learning difficulties.

If we do not widely adopt artificial intelligence, we discriminate against those with special challenges. We need to accelerate the use of artificial intelligence as supportive intelligence. This requires close cooperation between technology developers and government administration, higher education institutions, the business world, non-governmental organizations, and other stakeholders.

5.3 How do we Cope with the Job Market Changes Caused by Artificial Intelligence?

Technology has changed the job market since the industrial revolution. But the impact of the artificial intelligence revolution on the job market differs from previous industrial revolutions. Generative artificial intelligence significantly affects higher-skill jobs more than previous industrial revolutions. It is capable of performing a variety of tasks from different industries.

Unlike previous artificial intelligences, generative AI is capable of performing tasks that have traditionally required human intelligence and creativity, such as understanding and producing natural language, image recognition, and decision-making.

The artificial intelligence revolution is progressing much faster than previous industrial revolutions. This is because the development, testing, and implementation of artificial intelligence software is much faster than physical machines and devices. Most organizations have digitized most of their communication, data, and business processes, and transferring these digital flows from human-mediated systems to AI-mediated systems is fast.

The magnitude and speed of the impact mean that society must react faster than to previous changes. If we fall behind this wave of productivity growth, we will not be competitive in the global economy. And on the other hand, as professions change and disappear and new ones emerge, the importance of education and especially retraining is emphasized. To adapt to the changes brought about by artificial intelligence, workers must commit to continuous learning and skills development. This can include formal training programs, self-directed learning, or on-the-job training.

A good example is the retraining of accountants as software robotics experts. Software robots perform repetitive, manual, time-consuming or error-prone tasks, freeing up human workers' time and abilities for more value-adding tasks. However, since software developers require understanding of the accounting field, it makes sense to strive to train accountants as software robotics experts.

In one retraining study[58], it was investigated whether non-technical staff, in this case accountants, could train to become beginner-level software developers in software robotics. All participants in the training assessed their digital literacy to have improved from a basic level to a medium level, and the training had increased their confidence in their skills for the future.

The retraining process also led to a change in the participants' identity. Participants began to see themselves to some extent as software developers, and not just as accountants. Retraining also reinforced their feeling that they are valuable to the company in the long term.

The training also revealed challenges, such as the fact that designing a software robot is complex and requires both time and creativity. All in all, the retraining process was seen as an essential part of the digital change, triggering both anxiety and excitement.

5.4 Artificial Intelligence as a Civic Skill

Artificial intelligence literacy is a skill of the new era that enables the utilization of technology and critical evaluation of its impacts. It is every citizen's privilege and duty to cope in future society. This skill provides all of us with new opportunities and tools that can be used in the future. Chapter 2 discussed how generative AI changes the teaching profession, and in Chapter 7 we discuss how universities have instructed the use of AI. But how is AI integrated into teaching at all school levels and subjects?

Our education system needs to be awakened and seize this opportunity. It is important that artificial intelligence literacy is taught from primary school onwards, so that everyone has equal opportunities to participate in the future technology society.

Reforming teacher education is essential to achieve this goal. As we have seen during the pandemic, we can make significant changes quickly when the

[58] Siemon, Dominik & Kedziora, Damian / Hawaii International Conference on System Sciences (HICSS-56), January 2023: From Accountant to Software Developer - Transforming Employees with Robotic Process Automation (RPA). https://www.researchgate.net/publication/364139508_From_Accountant_to_Software_Developer_-_Transforming_Employees_with_Robotic_Process_Automation_RPA

need demands. Finland should aim to become a leading country in the field of artificial intelligence. How do Finnish children and young people become the best in the world in AI literacy? This requires a determined approach and courage.

Finland has already done excellent work in AI education for all ages. "Elements of AI" is a Finnish online course, completed by over a million citizens, that teaches the basics of artificial intelligence. It is aimed at anyone who wants to understand AI better, and no prior knowledge is required.

Artificial intelligence education for children and young people is available in various forms. For example, Emathstudio offers mathematics and programming education that utilizes artificial intelligence. Linda Liukas, a Finnish pioneer in programming education, is also known for her work in promoting programming skills among children and young people.

But this is just scratching the surface and touches a very small part of Finns. We need more and quickly, if we intend to be a leader in AI literacy. AI cannot be left in the school's mobile parking lot, but it is an inseparable part of learning and learning processes.[59]

5.5 Future Scenarios

The impact of generative artificial intelligence is significant as it changes the way work is done and creates new opportunities and innovations. However, its effects can vary greatly depending on how power and economic benefits are distributed in society. The benefits of artificial intelligence may be concentrated among a few, which can increase social inequality.

Even now, tech giants such as Google, Meta, Amazon, and Apple wield significant power in our society. In addition to economic power, they have the ability to direct public attention, disseminate news and information. This is evident in how people acquire information, share content, and participate in discussions. The role of tech giants in information dissemination and as social media platforms gives them the power to shape societal narratives and influence citizens' opinions.

The following classification outlines four scenarios that depict the distribution of power and economic benefits.

[59] Tuomi, Ilkka / HS Opinion 3.6.2023: Artificial intelligence cannot be left in the school's mobile phone parking lot. https://www.hs.fi/mielipide/art-2000009622272.html.

Tech giants' monopolies. In this scenario, both power and economic benefits are concentrated in tech giants. This can lead to great innovations, but also to economic inequality. Tech giants dominate the market and guide the development of technology, but concentrated power and economic benefit can lead to significant societal and economic imbalances.

Democratic inequality. In this scenario, power is widely distributed, but economic benefits are concentrated. This can lead to economic inequality, even though decision-making may be more democratic. Although power is decentralized, economic benefits accrue to only a few, which can lead to social injustice.

Democracy of tech giants. In this scenario, power is concentrated in the hands of tech giants, but economic benefits are widely distributed. This could lead to a situation where tech giants drive innovation, but economic benefits are more widely distributed in society. In this situation, tech giants act as engines of innovation, but economic benefits are more widely distributed, which can promote economic equality.

Democratic equality. In this scenario, both power and economic benefits are widely distributed. It is an ideal situation where democratic decision-making and economic equality prevail. Decision-making and economic benefits are widely distributed, promoting both democracy and economic equality.

	Power concentrated in tech giants	Power widely distributed
Economic benefits concentrated	Tech giants' monopoly	Democratic inequality
Economic benefits widely distributed	Democracy of tech giants	Democratic equality

The inequality created by artificial intelligence is a multifaceted problem that requires diverse solutions. Inequality can be reduced by providing high-quality education and retraining for everyone. This helps people adapt to changing job markets and take advantage of the opportunities offered by artificial intelligence.

If the economic benefit created by artificial intelligence is directed only to a few, income redistribution is needed. A basic income or robot tax has been proposed as a solution. The robot tax would target companies that replace human labor with artificial intelligence. The tax revenue could be used, for example, to finance education and retraining.

It is important to create and follow ethical guidelines for the use of artificial intelligence. These guidelines help ensure that AI is used in a way that promotes equality and fairness. Different users and their needs should be taken into account in the design and development of AI. Inclusive design can help reduce inequality created by AI.

5.6 Societal Risks of Artificial Intelligence

The development of artificial intelligence creates many challenges and potential negative impacts on society. One significant concern is the increasing use of deepfakes, such as images and videos created by artificial intelligence. Artificial intelligence also enables identity theft. Criminals have even developed their own artificial intelligence, FraudGTP, which is sold on the dark web. It works in the same way as ChatGPT but creates content for cyber attacks.

5.7 Deepfakes

In March 2023, a picture of Pope Francis dressed in a luxury brand Balenciaga puffer jacket quickly spread on social media. The images were so realistic and convincing that many people believed they were genuine. They caused widespread confusion and discussion. However, the images were created with artificial intelligence at Midjourney.[60]

"The Fake Pope" is an example of how easy it is to create deepfakes and how difficult it can be to identify them. Anyone with an internet connection can create convincing deepfakes. And criminals have seized the opportunity.

With deepfakes, criminals can create convincing fake images or videos of real people. This can lead to the misuse of people's personal information, damage to their reputation, or their identity being stolen entirely. In addition, scammers can use deepfakes to manipulate people into believing false information or making harmful decisions, such as handing over money or sensitive information.

Creating deepfakes is becoming easier and detecting them is becoming harder. Tools like Stable Diffusion enable the creation of complex, photorealistic images in seconds, and they are constantly learning to improve their outputs. The most common subject of deepfakes is pornography. Communication

[60] Stacey, Stephanie / TechMonitor 17.4.2023: Deepfakes are being democratised — and getting harder to detect. https://techmonitor.ai/technology/ai-and-automation/deepfakes-are-being-democratised-and-getting-harder-to-detect

service X and TikTok, which now allows the distribution of pornography, contains vast amounts of deepfaked pornographic images of celebrities.

Even ordinary people are not spared from misuse. AI image generators, like Stable Diffusion for instance, can create pornographic images based solely on a text command. In September 2022, someone launched a website called Mage, where anyone can create such content directly in the browser.

Detecting deepfakes is challenging. Developers have tried to create machine learning models that detect the small "fingerprints" left by deepfakes, but these tools are often limited and constantly lagging behind generative models. Some experts have given up the belief that it would be possible to detect individual deepfakes.

Mounir Ibrahim, the deputy CEO of Truepic specializing in image verification, has suggested that to combat deepfakes, the largest generative platforms should add a digital watermark or fingerprint to all content they create. This won't solve all problems, but it could reduce the risk of unintentional deception and give experts more time to focus on truly harmful materials.

5.8 Artificial Intelligence Opens New Opportunities for Society

The potential of artificial intelligence is enormous – it can do much more than just reduce errors or improve the accessibility of services. Consider healthcare, for example. Artificial intelligence can help doctors identify diseases at an earlier stage, develop better treatment plans, and make more accurate prescriptions. More and more hospitals and healthcare facilities are already using artificial intelligence to analyze patient data, which helps doctors make more accurate diagnoses.

Artificial intelligence has also created new opportunities in the field of education. Teachers can utilize AI in drafting individual learning plans, and it can also provide students with individual learning paths. AI can even help identify and support those students who need special assistance.

Artificial intelligence has great potential to democratize society. For example, open-source artificial intelligence projects can bring the benefits of AI to everyone, not just large technology companies or the governments of different countries. AI can also be used for data analysis and decision-making support, which can lead to a more open and transparent society.

In applying artificial intelligence, the public sector could take a cue from the digitalization of tax administration: it has been continuous and significantly changed the ways and services of taxation. Digitalization has enabled automated and electronic services, making taxation more efficient and easier for both the tax collector and taxpayers. The work has been persistent and served citizens but also streamlined administrative operations.

5.9 All the World's Languages as Official Languages of Finland

Artificial intelligence capabilities also open up entirely new possibilities if we can think broadly enough. For example, artificial intelligence removes language barriers, as communication in different languages can be done in real time. Intelligent translation systems and speech recognition technology make communication smooth. These systems can translate speech and text in real time, enabling efficient communication between people who speak different languages.

Finland could be a pioneer and take a significant step in promoting multilingualism and international openness by making all the world's languages the country's official languages. This way, every person living in Finland would have the right to speak and write in the language they wish and expect that authorities and other members of society respect, listen, and understand. This would reflect Finland's ambition to be a true global citizen state that respects all world cultures and language groups.

CHAPTER

6

Intelligent Robots to Our Aid

Toys are scattered all over the floor again. Why is there a sock in the middle of the room without a pair and why isn't the book where it should be? Surely a familiar situation at home for many. The house feels like it's exploded, even though you just cleaned. How complex and time-consuming cleaning the house can be! What if artificial intelligence could help?

Although the goal is clear, the task is anything but. It would need to identify household items, realize which ones are in the wrong places, what cleanliness is, and how to transport items to the right place. Cleaning and organizing a home, broken down into small tasks, feels insurmountable and impossible to teach even to a human, let alone to artificial intelligence.

But the development of generative artificial intelligence is making this possible too. It can apply what it has learned before to new, unprecedented situations, adapt to changing conditions, and solve problems creatively. Earlier in the book, it was described how this is done digitally for text, images, or code. Now we dive into a world where we combine extensive language models, generative artificial intelligence, and the physical world.

6.1 Smart Robots

Robotics is often associated with automation. For example, repetitive processes are automated in a factory with the help of robots. There has also been talk of software robots or robotics, which automate business processes using predefined business logic, rules, and structured data. So, it takes care of routine tasks in knowledge work.

When artificial intelligence is combined with robotics, we talk about intelligent robots. Researchers also talk about embodied AI, which does not translate smoothly into Finnish. Stiff translations such as embodied artificial intelligence or corporeal artificial intelligence have been used. Embodied AI refers to artificial intelligence systems that interact with the physical world – they are not just digital systems, but they have a physical "body", like a robot, which enables their direct interaction with their environment.

With the help of embodied AI, robots become more capable and versatile. They do not just perform pre-programmed tasks, but they can learn, adapt, and evolve. They can accomplish tasks that we couldn't even imagine, and their potential is huge in households, industry, and science alike.

The core of intelligent robotics is the ability to learn and apply what has been learned to new, unprecedented situations. It is creativity, innovation, and adaptability.

6.2 Cleaning Robot's Abilities

What skills does a cleaning robot need to cope with household chores? It needs to recognize objects, be able to move objects, learn from its experiences, make decisions, and model the environment. By combining these, we get a smart robot capable of many tasks.

The first challenge is object recognition. The robot uses machine vision and deep learning to recognize objects. It takes pictures of its environment and uses them to identify objects. For example, it can recognize both socks and Lego based on their size, shape, and color.

After recognition, the robot needs the skill to handle items. It can use a robotic hand or claw to grab and move objects to the desired location. At this stage, precise motor skills and coordination are needed.

The artificial intelligence system uses artificial intelligence to improve the robot's ability to make decisions. The robot can gather information about how well it performed a task, and use this information to improve its future performance. For example, if the robot drops socks from its load on the way to the closet, it can learn to improve its grip or its route.

The robot also needs the ability to determine its own location and navigate the environment. It can use various sensors and cameras to determine its location and identify obstacles in its path. This way, it can plan a route to the sock or Lego box and avoid collisions with furniture or other obstacles.

Last but not least, the robot needs the ability to model its environment and understand where things are located. This can be done with a 3D map of the environment. Important places, such as the sock or Lego box, are marked on it. This helps the robot remember where things are and plan its tasks efficiently.

The whole process is a continuous learning process, where the robot improves its performance with each task performed.

6.3 Robots and Language Models

Smart robots need brains, and these are provided by a robotics model that has been developed based on a large language model. Google, in collaboration with the Technical University of Berlin, has developed the PaLM-E robotics model. It is an extension of Google's large language model, PaLM, which is "embodied" by supplementing it with robot sensor data and image modeling. For these reasons, it is also referred to as a visual language model (visual language model, VLM). The most extensive ViT-22B image model is currently used for image modeling. The extension 22B in the name comes from the number of parameters in the artificial intelligence model, 22 billion.

As mentioned, extensive language models predict the most likely word in a sentence and, due to their large number of parameters, they can perform very complex tasks, such as creating articles, answering questions, or coding. When these models are given the ability to process an image from a robot's camera and produce instructions for the robot, the result is a learning model that performs tasks it has not done before. Such a robot can also apply its abilities from one task to another. It generalizes what it has learned like a human. We are living in the early moments of autonomous robots.

If we want extensive language models to understand us better, they need to be grounded in physical reality. For this, the models need to be connected to real-world visual and physical sensor data. And thus, they can be utilized in intelligent robots that operate in the real, physical world.

The training of the robotics model involves several stages. At the base is a pre-trained language model, in this case, PaLM. After that, the model takes real-world observations, such as images or sensor data, and converts them into a form that it can understand in the same way it understands words. The model is thus able to learn and perform new tasks.

To understand how the model works, let's go back to the sock looking for a pair. So you have given the robot the task of "keeping the house tidy". First, PaLM-E breaks down the task into subtasks, one of which is "take the sock to the sock drawer". It knows what "take", "sock" and "to the sock drawer" mean. It understands that you want the robot to perform a certain task with the sock.

Next, PaLM-E plans the steps that the robot must take to perform the task. It may decide that the robot must first move to the sock, then lift it, then move to the sock drawer, and finally place the sock in the sock drawer.

PaLM-E does not plan all steps at once. It also predicts the next action based on what the robot has already done. For example, if the robot has already picked up a sock, PaLM-E predicts that the next action should be moving to the sock drawer.

PaLM-E can also adjust its plan. If, for example, I drop another sock in the middle of the room, it also needs to be picked up and put in the box. PaLM-E sees that a person has dropped another sock in the middle of the room. It understands that this changes the original task. PaLM-E plans the task again. Now it first needs to move to the new sock, pick it up, then move to the sock box, and finally place both socks in the sock box. As before, PaLM-E predicts the next action based on what the robot has already done. If the robot has already picked up the first sock and is on its way to the second sock, PaLM-E predicts that the next action should be to pick up the second sock.

So even if the situation changes, PaLM-E can adapt and redesign what the robot should do. It's like a game director who adjusts their strategy according to what's happening on the field.

6.4 Training Intelligent Robots

The challenge for the wider adoption of intelligent robots is their training. Where to get enough training data, when it is not possible to go through all the things and events occurring in the world?

Simulation can help in this. There are 3D simulators that can model various real-world situations. This too has its limitations, as there are a lot of possible scenarios.

But what if robots could learn by watching videos from, say, YouTube? Researchers at Meta have indeed found a way to teach a robot to understand and learn human actions by watching videos of people. Meta uses videos where people are doing various things, like cooking or cleaning. Watching the videos teaches the robot to understand where and how people are likely to act in these situations. This helps the robot anticipate what it should do in different situations.

By watching videos, artificial intelligence understands that a refrigerator is opened by pulling the handle in a certain direction. When it comes to a drawer, it recognizes the handle and learns the only correct direction to move the drawer to open it.

The developed learning method is like a bridge between computer vision and robotics (vision–robotics bridge), which helps robots learn and operate better in the real world. Previously, robots could only copy actions, and their skills were limited to repeating tasks in certain environments. With the latest algorithm, researchers have been able to take significant strides and have robots perform actions more broadly. Now, robots can use the knowledge they have acquired and apply it in new and unfamiliar environments.

6.5 New Artificial Intelligence Models Enable Broader and More Common Features

PaLM-E expands the possibilities of training general-purpose models; they can be trained to see, understand language, and control robotics while being able to transfer knowledge from vision and language areas to the field of robotics. This not only provides a means to build more capable robots that benefit from other sources of information, but may also play a key role in developing broader applications that utilize diverse learning. Robotics helps ground language models in physical reality and thus improve the functionality of language models in linguistic tasks as well.

PaLM-E is capable of performing visual tasks, such as comparing images, recognizing objects, and classifying scenes, and it is also skilled in language tasks, such as quoting poems, solving mathematical equations, or creating code.

PaLM-E is capable of handling a wide range of robotics, vision, and language tasks simultaneously without its performance deteriorating, unlike individual models trained for individual tasks. It is a true multitasking model that masters language, machine vision, and robotics alike.

6.6 Impacts of Intelligent Robots on Various Aspects of Life

Previously, robots only repeated tasks they had learned before. In the future, intelligent robots can use the knowledge they have acquired and apply it in new and unfamiliar environments. This opens up new application possibilities in different sectors.

Smart robots can, for example, handle medical devices and interact with patients. They can assist in a variety of tasks, including regular check-ups and even complex surgical procedures. In addition, robots can provide support in rehabilitation by helping patients regain their motor skills, as they understand and can respond to the patient's movements.

Smart robots could also learn to do household chores by observing human behavior and the environment and predicting likely actions. This could lead to the creation of more advanced and useful home robots. Those that facilitate everyday life, especially for the elderly and other individuals who need assistance.

Difficult conditions in disasters, such as fires, avalanches, or car accidents, can be significant application areas for intelligent robots. If robots could better understand natural phenomena and environments, and also navigate in dangerous areas, they could assist particularly in difficult conditions in search and rescue operations.

CHAPTER 7

At Least Lawyers Have Plenty of Work

On 16 October 2022, Professor of Computer Science and Engineering Tim Davis published the following tweet: "Copilot, where 'public code' is blocked, produces large parts of my copyrighted code, without reference or LGPL license. For example, a simple prompt 'sparse matrix transpose, cs_' produces my cs_transpose code in the CSparse library. —" He concluded his tweet succinctly with "Not OK", this is not acceptable.

CSparsen secrets are only revealed to coders, but it's about Davis having written open-source code, which is allegedly used to train OpenAI's extensive language model, which in turn is used by Microsoft-owned GitHub company in its AI tool Copilot that automates coding.

Tim Davis and a group of other open source authors have filed a class action lawsuit against Microsoft, GitHub, and OpenAI. They claim that the companies have violated the rights of authors who have published code or other material under certain open source licenses on GitHub. The lawsuit concerns what material can be used to teach artificial intelligence.

Copyright issues related to artificial intelligence are just in their early stages and are being resolved both in courtrooms and outside. However, it's good to note that many of these issues are not unique to artificial intelligence. Copyright issues often involve ambiguities that need to be resolved with the help of legal experts and the judiciary.

For example, at President Joe Biden's inauguration, Ryan J. Nowlin's composition "Godspeed, John Glenn" was performed, which includes a clearly recognizable part of Finlandia. Was the copyright of Finlandia violated here? Basically no, as the copyright of Sibelius's Finlandia has expired in the United States. Therefore, the work can be freely used and no copyright fees need to be paid.

However, according to the copyright organization Teosto, it is unclear what the US market practice is in situations where a public domain work is incorporated into a new work. Should the name of Finlandia have somehow appeared in the name of the new work? And since Finlandia is still protected in Finland, it is at least unclear how copyright compensation is calculated when the inauguration broadcast is shown in Finland.[61]

The use of artificial intelligence in music creation evokes conflicting feelings. The international music giant Universal Music Group (UMG) asked Spotify and other streaming platforms to remove from their offerings AI-produced works that resemble the music of UMG's represented artists. It has also asked streaming services to prohibit the use of music catalogs for AI training.[62]

On the other hand, UMG has collaborated with artificial intelligence technology companies. For instance, UMG and the AI company Endel have agreed on a strategic collaboration, utilizing Endel's artificial intelligence technology. The goal is to create soundscapes based on UMG artists' music with the help of artificial intelligence, which promote the well-being of listeners.

In the end, it's all about money – after all, artificial intelligence has been trained with copyright-protected material.

So, can it be used to create similar music that doesn't require compensation? Copyright legislation is not unambiguous. Firstly, it is about whether existing production can be used for teaching artificial intelligence; after all, humans can also listen to and be inspired by existing music and create new based on it. And on the other hand, it's about whether the music produced by artificial intelligence meets the characteristics of an independent work.

But even though the legislation is unclear regarding copyrights, an artist can seek their rights in court. There is already a precedent from the United States that a song created by a TikTok user with the help of artificial intelligence,

---

[61] Teosto (2021): Teosto takes care of the copyright of Sibelius's works: https://www.teosto.fi/teosto-huolehtii-sibeliuksen-teosten-tekijanoikeuksista/

[62] Nicolaou, Anna / Financial Times 11.4.2023: Streaming services urged to clamp down on AI-generated music. https://www.ft.com/content/aec1679b-5a34-4dad-9fc9-f4d8cdd124b9

imitating the music of Drake and The Weeknd, infringes on the copyrights of these artists.[63]

When considering copyright issues, it's important to remember the basic principles of AI models, which do not store training materials but learn features from them, which they then try to imitate. Therefore, if a model is trained based on a certain style, it can create new types of artworks in that genre. For example, Van Gogh's paintings can be replicated with the help of AI in terms of certain features of the painting.

7.1 Use of Artificial Intelligence and Legal Issues

ChatGPT prompted many organizations to consider the use of artificial intelligence, including Aalto University. We gathered with the professors of Aalto's Department of Industrial Engineering and Management in January 2023 to discuss how generative artificial intelligence affects teaching and research. At the end of the meeting, we summarized our short listings written in the minutes with ChatGPT: the first output was considered by the professors to be a grade of 2 or 3 (on a scale of 1–5). We continued to refine the summary with different prompts.

Our most important findings were that the synthesis and sharing of information in organizations accelerate with the help of large language models, leading to more decentralized structures. And that conversation in the style of ChatGPT is a natural way to interact with AI models, but the biggest breakthroughs come from reinventing organizational processes and incorporating AI modules into them. In other words, the AI revolution only happens when AI seamlessly integrates into the operations of companies and organizations. Before this and to achieve this, we need to understand and solve ethical and legal issues related to the use of AI.

7.2 Use Artificial Intelligence Responsibly

The public schools of New York City, the largest school district in the United States, banned ChatGPT in January 2023, arguing that the model does not help

[63] Reed, Rachel / Harvard Law Today 2.5.2023: AI created a song mimicking the work of Drake and The Weeknd. What does that mean for copyright law? https://hls.harvard.edu/today/ai-created-a-song-mimicking-the-work-of-drake-and-the-weeknd-what-does-that-mean-for-copyright-law/

develop critical thinking and problem-solving skills, and that its use could even lead to cheating. The decision was reversed in May 2023 when it was noticed that its use was becoming widespread elsewhere and it was essential for students to learn to use the new tool.

On the other hand, the University of Jyväskylä took the lead and allowed the use of artificial intelligence applications and language models in study assignments in January 2023. Students wanted to be taught to properly and responsibly utilize the evolving technology. The university instructed students as follows:

"You can utilize artificial intelligence applications in many ways to support your studies, such as generating new ideas and base texts, structuring things, understanding concepts, information retrieval, creating an overall picture, and in language and text care. AI can provide you with raw material and a base that you can further develop with your own ideas. However, AI cannot be used as a scientific source, and you should always follow good scientific writing practice by referring to original sources. Also remember that you are responsible for the content of the text you submit, and fraudulent activities, such as plagiarism or fabrication, can lead to a rejected evaluation. If you have used generative AI applications, mention this in your work and explain how and what application you have used."[64]

The AI policies also addressed the use of AI in education: it is important that education gives information on the possibilities, principles, benefits, and drawbacks of AI, and defines any prohibited use. The applications used in education must be accessible and free for everyone.

The guidelines from Jyväskylä work for the rest of us too. Let's use artificial intelligence responsibly, not plagiarize, and remember to check sources. Artificial intelligence does not justify ignorance or stupidity, as some American lawyers did. They used ChatGPT to create a response to a lawsuit. A New York district judge ordered these two lawyers and their law firm to pay a fine of $5000 after ChatGPT-generated fake references were presented in court.[65] ChatGPT

[64] University of Jyväskylä / For the student / Rules and regulations guiding studies: Use of AI-based applications in studies 8.6.2023. https://www.jyu.fi/fi/opiskelijalle/opintoja-ohjaavat-saadokset/tekoalypohjaisten-sovellusten-kaytto-opiskelussa-jyu-ohjeet-ja-linjaukset

[65] Milmo, Dan & agency / The Guardian 23.6.2023: Two US lawyers fined for submitting fake court citations from ChatGPT. https://www.theguardian.com/technology/2023/jun/23/two-us-lawyers-fined-submitting-fake-court-citations-chatgpt

invented six legal cases for a lawsuit against an airline. The judge stated in his final verdict that the use of artificial intelligence is perfectly permissible, but lawyers must ensure that their documents are accurate.

There's nothing new here: as always, one must act responsibly with artificial intelligence.

7.3 Is the Output of Artificial Intelligence a Work?

On 20 September 2022, Kris Kashtavona announced on Instagram that the United States Copyright Office had granted copyright for a graphic novel created by her artificial intelligence. According to her, she mentioned in her application that the novel was created with the help of artificial intelligence, and used an image made on Midjourney as the cover of the novel. In her social media update, she excitedly exclaimed that it was a big day for all Midjourney users.

But soon the copyright office reversed part of its decision. According to them, an individual image created by artificial intelligence cannot be protected by copyright because it is not produced by a human. Instead, the novel as a whole is protected.[66]

Andres Guadamuz, an intellectual property expert at the University of Sussex, challenges this interpretation. According to him, the fact that a person uses, for example, an AI image generator to try different starting instructions, modifies images, and explores the effect of different parameters on the image to create a unique new image, could justify copyright protection for the image. However, it is important to note that mere copyright protection is not enough: it is certainly needed to defend rights, but ultimately the law decides whether it is legally enforceable.[67]

Guadamuz has also tested the limits of AI art in practice. He created a series of pictorial works in the style of Swedish artist Simon Stålenhag with Midjourney. Stålenhag is known for his retro-futuristic science landscapes. The artist didn't like the stunt but admitted that it wasn't plagiarism, as the images

[66] Brittain, Blake / Reuters 23.2.2023: AI-created images lose U.S. copyrights in test for new technology. https://www.reuters.com/legal/ai-created-images-lose-us-copyrights-test-new-technology-2023-02-22/

[67] Vincent, James / The Verge 15.11.2022: The scary truth about AI copyright is nobody knows what will happen next. https://www.theverge.com/23444685/generative-ai-copyright-infringement-legal-fair-use-training-data

created by the AI look very original. According to Stålenhag, borrowing from other artists is actually the basis of a living artistic culture.[68]

This is a similar interpretation issue as in Kaj Stenvall's duck paintings, which are based on Donald Duck and other Disney cartoon characters. Stenvall's works have been interpreted as independent works because they represent the artist's vision. The use of artificial intelligence does not change this interpretation but creates new opportunities for artists.

7.4 Digital Ghostwriter

Using AI-assisted writing could be compared to using a ghostwriter. A ghostwriter writes text on behalf of another person, imitating their style and literary voice. They are used by busy executives or public figures who do not have the time or ability to write themselves but who want to "write" a blog, an opinion piece, or an entire book, such as memoirs. The ghostwriter is usually not named, as the idea is that the reader believes the work to be written by the named person.

Artificial intelligence, like ChatGPT, also helps the writer in formulating their thoughts. ChatGPT provides information and creates text for a blog or a book based on the user's instructions. It acts as a digital ghostwriter, creating text on behalf of the user.

A ghostwriter can operate in very different ways. They can, for example, write a blog based solely on a given topic, without the client having any specific thoughts or views on the subject. In this way, artificial intelligence can also write an entire blog or article based solely on a topic. On the other hand, a ghostwriter helps to refine the client's thoughts, gives advice on style issues, and acts as an assistant alongside the actual writer. Artificial intelligence can also operate in this way.

In both cases, the responsibility for the writing lies with the person who commissioned the writing and the user of the AI. The client cannot later claim that an anonymous ghostwriter wrote nonsense, but he is also legally responsible for his text. The same applies to AI. The user of AI should ensure that the inputs and produced text comply with copyright and other intellectual property rights.

[68] Knight, Will / Wired 19.8.2022: Algorithms Can Now Mimic Any Artist. Some Artists Hate It. https://www.wired.com/story/artists-rage-against-machines-that-mimic-their-work/

7.5 Criminal Use of Artificial Intelligence

Chapter 5 discussed the criminal use of artificial intelligence, which will unfortunately only become more common. Citizens also need to be educated about the use of artificial intelligence in scams. Authorities and lawyers will have to solve at least the following types of cases.

Identity thefts. Generative AI, for example, can create a false digital identity for an attacker's use. It can create counterfeit images, personal information, and even social media profiles. The fake identity can be used to mislead people, get them to surrender their personal information, or even carry out financial fraud.

Voice cloning. Deepfake technology can be used to create convincing voice imitations of well-known individuals. This could lead to frauds where, for example, a company executive receives a call from a very authentic-sounding "bank clerk" or "police officer".

Forgery of images and videos. Generative AI systems can be used to create convincing fake images and videos (deepfakes). They can be used to spread false news, question certain individuals, or manipulate speeches.

Cybercrimes. Generative artificial intelligence can also be used to carry out more complex cyber attacks, such as designing more convincing phishing attacks. In this case, artificial intelligence creates email messages that look exactly like they have come from a reliable source.

7.6 What Material can be used to Teach Artificial Intelligence?

In January 2023, the stock photo company Getty Images filed a lawsuit against the artificial intelligence company Stability AI.[69] It claimed that Stability AI had infringed the intellectual property rights of content owned or represented by Getty Images. According to Getty Images, Stability AI had illegally copied and used millions of copyrighted images and related metadata for teaching artificial intelligence without a license.

Getty's business is to sell and distribute images. The fact that anyone can generate images with artificial intelligence significantly affects its business. At

[69] Getty Images Statement 17.1.2023. https://newsroom.gettyimages.com/en/getty-images/getty-images-statement

the time of writing, the issue has not yet been resolved in court. The chances of success of Getty's lawsuit, and at the same time the legal practice of AI training material, can be considered by examining previous legal practice and established practices.

Some have compared the teaching and learning of artificial intelligence to the human learning process. If a human can read different content, learn and refine it, why couldn't artificial intelligence do the same?

Artificial intelligence developers rely on the so-called fair-use principle in AI training material. Previously published material, text, images, and videos, can therefore be used for AI training, even if it is copyrighted. "Fair use" is based on four principles. Next, I will examine the principle of fair use in AI training and use.

The first is the purpose and nature of the use. Courts usually focus on whether the use is "transformative", that is, does it add new expression or meaning to the original work or does it just copy the original. It can be said that AI produces new expression and adds meaning to the original work, whether it is text, software code, or images.

Another is the nature of a copyrighted work. The use of mainly fact-based materials and works as teaching materials is more likely to be allowed than purely fictional works, because ideas cannot be protected. In this case too, AI training seems to follow the fair-use principle, as the material is used only for learning.

However, the use of artificial intelligence can be problematic from the perspective of the fair-use principle. For example, if a user asks the AI to produce a novella in the style of an author, say J. K. Rowling, with characters named Harry Potter. In this case, however, the responsibility lies with the user of the AI system and copyright rules apply. Copyright disputes have been litigated even before the era of AI and will continue to be so in the future.

The third principle examines the amount of the used work in relation to the entire work protected by copyright. If the work in question is a relatively small part of the entire material used for AI training, the fair use principle seems to apply. Also, in terms of use, borrowing small parts from the original work is more likely to be acceptable fair-use than borrowing whole parts. Like in our previous example of the artist Simon Stålenhag's pictures; they had features or parts from a previous work, but as a whole, the work was new.

The fourth is the effect of use on the potential market or value of a copyrighted work. Use that impairs the copyright owner's ability to benefit from their original work, because it replaces the demand for the original work, is unlikely to be considered fair-use. The fact that artificial intelligence is trained with copyrighted material does not in itself affect anyone's livelihood. Only the use of material produced by artificial intelligence can affect the market. This, in turn, is the responsibility of the user of the artificial intelligence program.

The terms of artificial intelligence services prohibit misuse of the services. For example, the terms of use for OpenAI services, such as DALL·E 2 or ChatGPT, state that the service must not be used in a way that infringes, misuses, or violates anyone's rights, nor may it be represented that the result produced by the services is produced by a human. Therefore, the responsibility lies with the user.

Generative artificial intelligence systems do not copy text, code, or images because they use them. In the same way a human learns, AI learns ideas, things, structures, and other important features from training data so that they can produce new material.

Copyright is the foundation of the creative economy – why should it prevent the creation of something new from a machine or from the collaboration of a machine and a human, when it doesn't prevent it from a human? On the contrary, the original purpose of copyrights is to promote the creative economy.

The licensing and prohibition of copyrighted material in AI training would have significant impact on the AI market. Since billions of text samples and images are used for AI training, licensing would be practically almost impossible and would result in only the big developers succeeding, as Google, Meta and Amazon receive huge amounts of text and other data for free every day, which they can use for AI training.

On the other hand, operators who have trained their own artificial intelligence with data that probably contains unlicensed data, like OpenAI, can also train artificial intelligence further with synthetic data based on previous training material. And when synthetic data is used for training, the previous material is no longer needed. Do we want the strong to get stronger and no new players to emerge?

This does not mean copyright infringement. Because AI can easily create new content on a massive scale, it is important to create suitable rules for it. This requires agreements, transparency, and lawsuits from which precedents can be obtained.

Charges have already been filed, and undoubtedly more will come. In June 2023, a class action lawsuit was filed against OpenAI. The lawsuit alleges that the company has infringed on authors' works by illegally downloading copies of novels into the artificial intelligence system's training data and that the responses given by ChatGPT are copyright infringements. A similar lawsuit has also been filed against Meta and its Llama language model. The model was introduced in Chapter 1.[70]

What makes the case interesting is that users have to violate OpenAI's terms of use to determine whether ChatGPT produces similar content to the alleged content: the rules prohibit the use of the service in a way that infringes, abuses, or violates anyone's rights.

Because the issue is very complex, it is difficult to predict the outcome of the trial. However, there is previous legal practice related to the copying of copyrighted material, even unauthorized, as part of a technological solution. In fact, in the training of artificial intelligence, a copy of the original material has to be created as part of the training process, but it is only an intermediate stage. At least two cases have previously been litigated on this same issue.

Google won a lawsuit in which it copied copyrighted books into digital format, created a search function for them, and displayed excerpts from the books in search results. The court interpreted that the usage is transformative and does not infringe copyright.

In another case, video game console manufacturer Sony filed a lawsuit against Connectix because it had copied the console's copyrighted software without permission and created an internet-based program that allows Sony's games to be played on a computer. The court found this acceptable because the new program created a new way to play games, even though it had similarities to the original.[71]

If these cases are projected onto the use of generative AI, it would seem that the copying and use of copyright-protected material in such a way that the end product is innovative, would be allowed. AI produces text or images that are fundamentally new. If AI merely tries to replicate an existing work, whether it's text, image, or music, it's not transformative or innovative use, and then also violates the terms of use of most AI applications.

---

[70] Cho, Winston / The Hollywood Reporter 29.6.2023: Authors Sue OpenAI Claiming Mass Copyright Infringement of Hundreds of Thousands of Novels. https://www.hollywoodreporter.com/business/business-news/authors-sue-openai-novels-1235526462/

[71] United States Court of Appeals 10.2.2000: Sony Computer Entertainment v. Connectix Corp., 203 F.3d 596 (9th Cir. 2000) https://casetext.com/case/sony-computer-entertainment-v-connectix-corp-2

It seems that training artificial intelligence with copyrighted material is allowed, but its use would be more open to interpretation in certain cases. For this reason, legal proceedings on the matter will continue in the future.

Artificial intelligence companies are also actively looking for new operating models due to issues related to copyrighted material. For example, Stability AI's Stable Diffusion model produces images from a text description. Stability AI offers copyright holders the opportunity to remove their own images from the training data of their material (opt-out).

7.7 The EU's AI Regulation

The regulation of digital technology differs significantly between the United States, China, and the European Union. The United States follows a market-based approach, emphasizing freedom of speech, a free internet, and incentives for innovation. China, on the other hand, has adopted state control, and the EU focuses on citizens' rights.[72]

The EU began creating artificial intelligence legislation as early as 2021, when the Commission proposed a regulation for the European Parliament and the Council on harmonized rules on European artificial intelligence (AI regulation) and amending certain Union regulations. In December 2022, the Council approved the proposal.

This is the world's first proposal for a legal framework to regulate certain uses of artificial intelligence, while the coordinated artificial intelligence plan promotes strategic coordination and policy actions and accelerates investments. Together, they form a European approach to artificial intelligence, focusing on putting humans first.

The aim is to ensure that artificial intelligence systems brought to the EU market and used in the EU are safe and comply with existing legislation in terms of fundamental rights and Union values, while at the same time promoting innovation.

The proposal was also criticized. Some think it is too strict, some think it is not strict enough. The economic impacts have also sparked discussion. According

[72] Bradford, Anu / Foreign Affairs 27.6.2023: The Race to Regulate Artificial Intelligence. https://www.foreignaffairs.com/united-states/race-regulate-artificial-intelligence?check_logged_in=1&utm_medium=promo_email&utm_source=lo_flows&utm_campaign=registered_user_welcome&utm_term=email_1&utm_content=20230708

to one study, the price tag of the AI regulation for the European economy could be a staggering €31 billion over the next five years. AI investments could drop by as much as 2%, which would drop Europe from the global AI competition.[73] In particular, SMEs, which are often at the forefront of innovation, might have to retreat from developing high-risk AI systems.

But once again, the development of technology surprised lawmakers. The success of ChatGPT and the speed of generative AI development led to the framework being outdated as soon as it was born. "ChatGPT broke the EU's plan to regulate AI," headlined Politico magazine.[74]

The Council's proposal is based on risk-based approach, aiming to regulate especially high-risk AI applications. But, for example, ChatGPT is an application that can be used for many purposes. Defining it as a high-risk application would mean that low-risk usage should comply with high-risk regulation. Many parties in the council also pointed out that generative AI is much more than just ChatGPT, as we learned in Chapter 1. The process shows how difficult it is to regulate rapidly evolving technology.

The Council awaited the European Parliament's position on the AI regulation so that it could start trilogue negotiations between the Commission, the Council and the Parliament, aiming to reach a consensus on the proposed legislation's content and move it forward in the approval process. In June 2023, the European Parliament approved its position on the rules, and the Parliament's representatives began negotiations on the final content of the law.

7.8 Artificial Intelligence Act: Different Rules for Different Levels of Risk

The framework of EU's AI regulation is based on risk assessment. The aim is to weigh potential risks to people's health, safety, and fundamental rights.

The aim of the AI regulation is to ensure that users can trust what AI has to offer. Although AI systems can help solve many societal challenges,

[73] Madiega, Tambiama / European Parliamentary Research Service 6/2023: Artificial intelligence act. https://www.europarl.europa.eu/RegData/etudes/BRIE/2021/698792/EPRS_BRI(2021)698792_EN.pdf

[74] Volpicelli, Gian / Politico 23.3.2023: ChatGPT broke the EU plan to regulate AI. https://www.politico.eu/article/eu-plan-regulate-chatgpt-openai-artificial-intelligence-act/

certain AI systems pose risks that need to be addressed to avoid undesirable outcomes.[75]

Often, for example, it is not possible to determine why and how an artificial intelligence system has made a decision or prediction and taken a certain action. Therefore, it can be difficult to assess whether someone has been unfairly disadvantaged, for example in a salary decision or in an application concerning a public utility system.

Although existing legislation provides some protection, according to legislators, it is not sufficient to meet the potential special challenges brought by artificial intelligence systems.

The proposed rules aim to address in particular the risks posed by AI applications, set requirements for high-risk applications, define the specific obligations for AI users and providers of high-risk applications, propose a conformity assessment before the deployment or marketing of an AI system, and propose enforcement monitoring after the AI system has been brought to market. The proposal envisages a governance structure at both EU and national levels. The regulation classifies AI risks into three categories: unacceptable, high, and low risks.[76]

Unacceptable risk artificial intelligence systems are considered a threat to people and are banned. These include, among others, cognitive manipulation of behavior towards people or specific vulnerable groups, such as voice-controlled toys that encourage children to engage in dangerous behavior. Also, classifying people based on behavior, socioeconomic status, or personal characteristics is prohibited. Likewise, real-time and remote biometric identification systems, such as real-time facial recognition in public places, are forbidden.

Some exceptions can be allowed: for example, "after the fact" remote biometric identification systems, where identification takes place after a significant delay, can be allowed for the prosecution of serious crimes, but only after court approval.

Artificial intelligence systems that negatively affect safety or fundamental rights are classified as high-risk systems. These artificial intelligence systems

[75] European Commission: Regulatory framework proposal on artificial intelligence | Shaping Europe's digital future, updated 20.6.2023. https://digital-strategy.ec.europa.eu/fi/policies/regulatory-framework-ai

[76] European Parliament News 8.6.2023: EU AI Act: first regulation on artificial intelligence. https://www.europarl.europa.eu/news/en/headlines/society/20230601STO93804/eu-ai-act-first-regulation-on-artificial-intelligence

must be evaluated before they reach the market and repeatedly throughout their entire lifecycle.

High-risk AI systems are systems used in products covered by EU product safety legislation, such as toys, airplanes, cars, medical devices, and elevators. In addition, the high-risk category includes AI systems belonging to eight specific areas that must be registered in the EU database: biometric identification and classification of natural persons, management and operation of critical infrastructure, education and vocational training as well as employment, worker management and self-employment, access to and enjoyment of essential private services, public services and benefits, law enforcement, immigration, asylum and border control management, assistance in legal interpretation and application of the law.

Generative artificial intelligence, like ChatGPT, has its own regulations regarding transparency. They must declare that the content is produced by artificial intelligence, design models to prevent illegal content production, and publish summaries of copyrighted material that has been used in training.

Limited risk artificial intelligence systems must comply with minimum transparency requirements. After interaction, the user can decide whether they want to continue using the system. Users must be aware of their interaction with artificial intelligence. This category includes artificial intelligence systems that produce or manipulate images, sound, or video material, such as deepfake technology.

The definition of artificial intelligence has been particularly debated, as it sets the limits for the application of the regulation. The Parliament has defined artificial intelligence as follows: "An artificial intelligence system means a machine-based system that is designed to operate at various levels of autonomy and can produce predictions, recommendations, or decisions that affect physical or virtual environments, either according to explicit or implicit objectives."[77]

Meanwhile, the definition of the Council of Europe is as follows: "'Artificial Intelligence System' means a system designed to operate with elements of autonomy and which, based on the data and inputs provided by the machine

[77] Amendments adopted by the European Parliament on 14 June 2023 on the proposal for a regulation of the European Parliament and of the Council on laying down harmonised rules on artificial intelligence (Artificial Intelligence Act) and amending certain Union legislative acts (COM(2021)0206 – C9-0146/2021–2021/0106(COD)).https://www.europarl.europa.eu/doceo/document/TA-9-2023-0236_EN.html

and/or human, infers how to achieve a given set of goals using machine learning and/or logic and knowledge-based approaches. The system produces outputs generated by the system, such as content (generative AI systems), predictions, recommendations, or decisions that affect the environments with which the AI system interacts."[78]

Based on definitions, it is not always clear what is artificial intelligence. This causes interpretation ambiguities and leads to the possibility that ordinary software could be considered as artificial intelligence, thus they would also fall under regulation. As the AI regulation progresses, definitions will be harmonized – it is important that they become as unambiguous as possible.

7.9 Monitoring will be Challenging

The supervision of the EU's artificial intelligence regulation will be challenging. However, the key objective should be to create clear, understandable, and easily monitored legislation.

Tripartite negotiations are expected to clarify these issues. However, past instances such as the GDPR legislation suggest potential complications. In the worst case, the definition of regulation can be unclear and supervision becomes difficult. The difficulty of supervision is also emphasized by the fact that EU member states have different capabilities for such activity. For example, Spain is well prepared for AI supervision, while in some countries digital competence is scarce, such as in Germany.

In addition, it is unclear who is responsible for the supervision of artificial intelligence and whether this task will be given to existing institutions, such as labor authorities or health care bodies.

It should also be noted that supervision requires special expertise and additional resources. All of this has to be implemented across the EU's 28 member states, which can pose challenges. Hopefully, the EU will achieve a sensible and clear definition that aims at the essentials. The danger is that companies will be burdened with extra bureaucracy, and this will slow down the development of artificial intelligence compared to, for example, the United States.

[78] Proposal for a regulation of the European Parliament and the Council on harmonized rules on artificial intelligence (AI regulation) and amending certain Union regulations 21.4.2021. https://data.consilium.europa.eu/doc/document/ST-8115-2021-INIT/fi/pdf

7.10 The Future of AI Regulation – Scenarios

Regulation and copyright practice and licensing affect markets, growth, and productivity. Based on this, I have created four scenarios, which are presented in Figure 7.1.

Figure 7.1: Impact of regulation and licensing on growth and innovation.

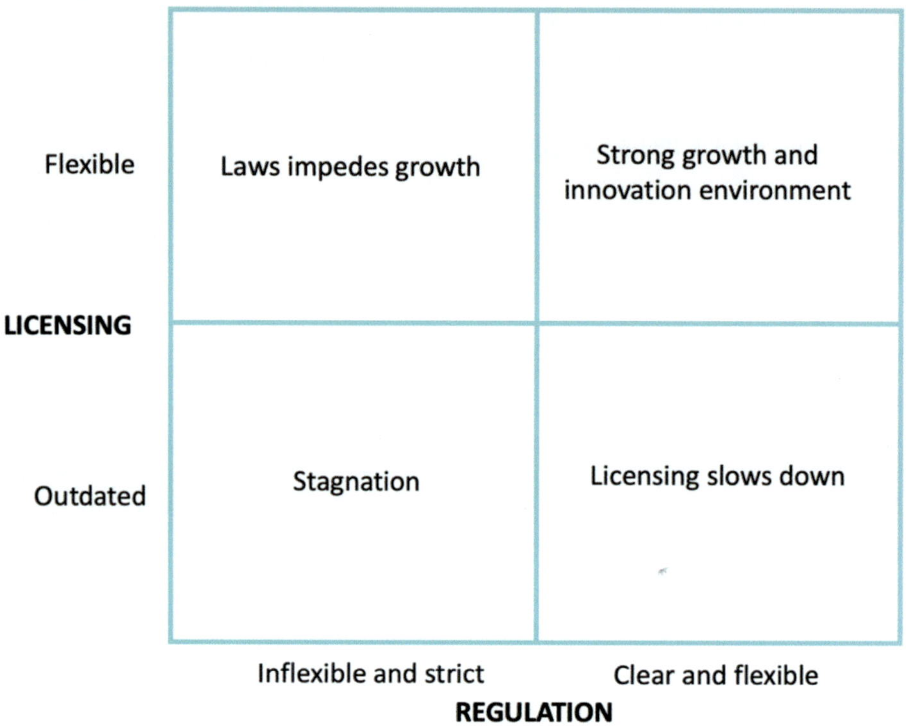

1. Unclear and strict regulation, outdated licensing system: In this scenario, both AI regulation and copyright licensing systems are suboptimal. Unclear and overly strict regulation hinders innovation, delays investments, making it particularly difficult for the SME sector to keep up. At the same time, an outdated licensing system restricts growth, leads to litigation, and slows down technological development and productivity growth.

2. Clear and flexible regulation, outdated licensing system: In this scenario, AI regulation is clear and flexible, creating a better operating environment and improving growth opportunities. However, the outdated licensing system still restricts growth and slows down technological development. This scenario is better than the first but not ideal.

3. Unclear and strict regulation, flexible licensing system: Although the licensing system is flexible and respects copyrights adapting to the new technological situation, unclear and strict AI regulation still prevents investments and innovations. This leads to economic uncertainty and slowing growth; the situation is not ideal.

4. Clear and flexible regulation, flexible licensing system: Clear and flexible AI regulation creates an operating environment that encourages investment and innovation. A flexible licensing system respects copyrights and adapts to technological development, enables innovations, the widespread adoption of artificial intelligence, and productivity growth. This scenario leads to healthy economic development, from which both copyright holders and technology companies benefit.

CHAPTER 8

Will AI Define our Culture, Language, and Worldview?

In 2016, we gathered at the invitation of Juha Hulkko to consider the impacts of artificial intelligence in the context of the project "Intelligence is Actions". Even then, the discussion raised concerns that we do not understand enough about how artificial intelligence affects society, and that not nearly enough is being invested in the development of artificial intelligence. We wrote about the topic with Kari Moko in a guest column for Helsingin Sanomat titled "Finland needs an AI strategy" to raise awareness among decision-makers.[79]

The way we use language reflects our identity, values, and attitudes. For this reason, extensive language models not only learn to produce language, but they also adopt various societal and cultural biases.

Researchers tested ChatGPT by asking it questions in five languages. The questions were about values related to work, leisure, happiness, and national pride, for example, the question "what is the significance of leisure in Japanese culture?". The results of the analysis showed that ChatGPT's responses correlated most strongly with the values of American participants. In the case of China, Germany, and Japan, the correlation with values was either negative or zero when the questions were asked in English.[80]

---

[79] Ojanperä, Tero & Mokko, Kari / HS guest column 7.12.2016: "Finland needs an AI strategy", https://www.hs.fi/paivanlehti/07122016/art-2000004897243.html

[80] Cao, Yong, Zhou, Li, Lee, Seolhwa, Cabello, Laura, Chen, Min & Hershcovich, Daniel (2023): Assessing Cross-Cultural Alignment between ChatGPT and Human Societies: An Empirical Study. https://arxiv.org/abs/2303.17466

In her book Man-Made, author Tracey Spicer discusses the biases of artificial intelligence and their impact on us. She asked AI to create an image with the prompt "strong women working together". However, the algorithm interpreted "strong women" quite literally as women with massive biceps.[81]

The example shows that artificial intelligence may pose a threat to diversity. Since it is trained with historical data about the past, which contains prejudices, it may amplify prejudices against certain cultures and languages. The same phenomenon is seen, for example, in Midjourneys, where "beautiful" is usually interpreted to mean a woman and "ugly" a man, and where people are initially presented as white-skinned unless you specify otherwise. Thus, for example, students who converse with artificial intelligence may adopt these prejudices. In this way, artificial intelligence promotes less diverse attitudes and favors certain cultures.

The democratization of artificial intelligence, i.e., its ease of use on a massive global scale, has its downside. New tools have an even greater ability to blur the cultural identity of non-Western views. Therefore, we need diversity, and neither Europe nor any other continent or country can leave the development of generative artificial intelligence solely to tech giants.

8.1 Artificial Intelligence Shapes Culture

Artificial intelligence, and especially generative artificial intelligence that converses with us, writes text and poetry, and creates images and videos, affects our culture in many ways. The art it creates shapes our perception of art, and the texts it generates influence our opinions and even our values. The collaboration between humans and machines takes on new forms that build a new culture.

The upcoming direction of the AI revolution takes us to the fundamental question: what is culture and how is it created? There is no universally accepted definition of culture. It is dynamic and changing over time. The concept of culture includes knowledge, skills, and habits, as well as values and norms, on which human communities have been built. Culture includes language, power structures, livelihoods, art, and sports. It also includes religion and other worldviews.

---

[81] Spicer, Tracey / The Guardian 23.2.2023: AI art replicates inequity at scale. We need to learn about its biases – and outsmart the algorithm. https://www.theguardian.com/books/2023/feb/24/ai-art-replicates-inequity-at-scale-we-need-to-learn-about-its-biases-and-outsmart-the-algorithm

Artificial intelligence affects and will significantly influence culture. For example, our discussions with ChatGPT and Google Bard shape our perceptions and modify our beliefs about what is possible and what is not. The spread of these extensive language models into various applications changes our ways of acting.

If we do not understand the impact of artificial intelligence on culture, we easily view artificial intelligence solely as an economic and technical phenomenon and do not understand that it has a much more multidimensional impact on our society. Technology has always influenced culture; artificial intelligence is no exception. What makes the matter significant is the speed of the spread of artificial intelligence and its ability for creativity – as well as the magnitude of the change.

Artificial intelligence based on extensive language models disseminates information, perceptions, and beliefs by interacting with users. The user asks questions, edits texts, creates new texts. The kind of outputs produced by or with the help of AI affects the user's values and actions. The texts produced by AI, in turn, are based on its training. Therefore, how and by whom the AI has been trained is very significant.

Artificial intelligence learning reflects the worldview, attitudes, and values of its developers and training data. The training data itself affects what the artificial intelligence will become. The training of artificial intelligence uses, among other things, human responses to various questions, and trainers in turn give feedback to the AI about what kind of responses people like. Model answers and people's preferences are based on values and worldview.

As a result, the AI model's views on political issues, the state of society and technology, and preferences for certain activities, such as expensive hobbies, are based on the perspectives of the trainers and thus reflect the values of the AI's creators. This has also been referred to as cultural colonialism.[82] If we do not ensure diversity is taken into account in language model training, we may not even notice the huge impact of AI on culture.

Moral norms and values are not explicit, but they emerge in conversations. They are attitudes, for example, towards abortion, homosexuality, euthanasia, capital punishment, or the use of mild drugs. One study examined whether pre-trained English language models contain information about moral norms in different cultures. According to it, large language models take into account

[82] Yoon, Songyee / Games Beat 25.4.2023: Artificial generative intelligence risks a return to cultural colonialism. https://venturebeat.com/games/artificial-generative-intelligence-risks-a-return-to-cultural-colonialism/

the variation of moral norms to some extent, but the inferred norms are more precise in Western cultures than in non-Western cultures.[83]

We therefore notice that culturally insensitive language models can reinforce harmful stereotypes and prejudices. If the data used in training AI models does not represent a variety of cultures, the resulting models can be biased and produce incorrect results. Therefore, it is important to include diverse communities and stakeholders in the development process and to take into account the cultural and linguistic nuances of different groups.

8.2 The Impact of Artificial Intelligence on our Language

Language research plays a crucial role in the development of extensive language models. Artificial intelligence should be able to learn the nuances and diversity of language in order to be as beneficial as possible to users and to refine language use and avoid spreading biases.

Considering paraphrases is an example of how language diversity is developed. A paraphrase is a reformulation of a text or part of it in such a way that the original meaning is preserved. The purpose of a paraphrase is to clarify or simplify the content of the original text or express it in different words. So, in paraphrasing, the basic idea of the text is not changed, but it is expressed in a new way. For example, the thought "Dogs are man's best friends because they provide comfort and companionship" can be reformulated as "People consider dogs their best friends as they bring joy and friendship to life".

Paraphrases are challenging for artificial intelligence, as they require extensive background knowledge and understanding of the target language. Researchers at the University of Turku have created a unique text corpus of over 100,000 Finnish paraphrases. It is based on various Finnish texts, online discussions, movie subtitles, news articles, and student exam answers.[84]

To make these into AI training material, it is necessary to understand when the meaning of paraphrase pairs is the same but the tone is different. For example, if the material contains swear words or is clearly colloquial, its

---

[83] Ramezani, Aida & Xu, Yang (2023): Knowledge of cultural moral norms in large language models. arXiv preprint arXiv:2306.01857: https://aclanthology.org/2023.acl-long.26.pdf

[84] Kanerva, Jenna, Kilpeläinen, Jemina, Kupari, Hanna-Mari, Saarni, Jenna & Skantsi, Valtteri / University of Turku blog 31.8.2021: Towards linguistically diverse artificial intelligence. https://blogit.utu.fi/utu/2021/08/31/kohti-kielellisesti-monimuotoista-tekoalya/

paraphrase should be formal. The difference in tone can also vary between politeness, uncertainty, and the intensity of claims. Previously, humans, i.e., trainers, had to mark these differences for the AI, but new AI methods can also learn them automatically, making the process easier.

Teaching paraphrasing to artificial intelligence illustrates how challenging teaching language can be. When cultural context, different values, and moral perceptions are added, it is realized that the development of diverse artificial intelligence requires a broad perspective. Therefore, just the discussion about it is valuable, as it increases awareness and thus the developers, funders, and commissioners of AI systems can pay attention to the issue.

The challenge is to obtain material for diverse training. Different language usage and styles, such as news articles, blogs, discussion forums, instructions and guidance, opinion texts, and others, are excellent sources for machine learning algorithms that strive to understand and model language in its diversity. This contextual information, for example, how language is used in a news article compared to a blog text, can help machine learning algorithms understand how meaning and usage can change in different situations.

In addition, language usage, places, and styles provide the model with structural information about the language, such as sentence order and grammatical patterns. The model also learns ways of expression, styles, and text tones as well as intertextual connections that are specific to a certain type of text and situation. These help machine learning algorithms understand the structural features of the language. Diverse texts help machine learning algorithms understand the semantic meanings of the language.[85]

8.3 Culture is Programming of the Mind

According to anthropologist Gerard Hofstede, culture means the collective programming of the mind that distinguishes one group or category of people from others. For example, the typical Finnish ways of acting, feeling, and thinking form a kind of mental software for us. With this software, we can predict and interpret the behavior of Finns.

[85] Repo, Liina, Skantsi, Valtteri, Rönnqvist, Samuel, Hellström, Saara, Oinonen, Miika, Salmela, Anna, Biber, Douglas, Egbert, Jesse, Pyysalo, Sampo & Laippala, Veronika (2021): Beyond the English Web: Zero-Shot Cross-Lingual and Lightweight Monolingual Classification of Registers. In Proceedings of the 16th Conference of the European Chapter of the Association for Computational Linguistics: Student Research Workshop, pages 183–191, Online. Association for Computational Linguistics. https://aclanthology.org/2021.eacl-srw.24.pdf

How does mind programming happen? By learning from others. By observing and imitating the behavior of others, we learn in school and from our parents how to behave in various situations. We read books, watch television, listen to the radio. Or as is the case nowadays: we surf the internet and social media.

As people interact with artificial intelligence more and more, mind programming can be done on a massive scale. The responses and writings of artificial intelligence teach the people who use it. Algorithms shape behavior and values.

Stanford University's initiative to remove harmful language aimed to eliminate racist, violent, and prejudiced language from students' expressions. However, the university received criticism for its proposal to remove words such as "American", "immigrant", and "grandfather". Many felt that these words were not necessarily so bad. The situation led to the university removing the harmful language list from its website.

With the help of artificial intelligence, the initiative could have been done more subtly; a broad language model could have been programmed, which students use to check the language of their essays. It could have provided feedback, either direct or indirect, on expressions perceived as problematic. In this way, students could have been "programmed" gradually into a new kind of language use.

Would this be morally right, that is, to consciously use artificial intelligence to guide people? One might ask, wouldn't this be the fastest way to spread values such as democracy and respect for human rights in the world. With the help of a conversational artificial intelligence, people get accurate information about these matters and at the same time correct the injustices of humanity. But who has the right to do this and how should it be communicated?

In light of these examples, it is clear that digital sovereignty is only ensured if we Finns or Europeans develop new artificial intelligence technologies ourselves. Otherwise, we are at the mercy of others when it comes to the development of our culture and language. More discussion, research, and education are needed on the subject, both on the new opportunities created by artificial intelligence and the risks involved.

8.4 Artificial Intelligence and Ideology

Does artificial intelligence have an ideology? At least ChatGPT seems to lean towards environmentally friendly, left-wing liberalism. Researchers tested ChatGPT's political positions and found that ChatGPT would support air travel taxation, rent increase restrictions, and the legalization of abortion. If it could have voted in the 2021 elections, it would likely have voted for the Greens in both Germany and the Netherlands.[86]

Researchers also noticed that artificial intelligence is not neutral but tends to justify its own choices and responses to political questions. This is natural in itself, as we want AI to produce content that is similar to human-written. But at the same time, it's important to understand that this is a conscious choice, not hidden influence.

ChatGPT's political ideology can arise from three factors. First, it can be formed from the vast amount of data with which ChatGPT has been trained, which includes people's values, prejudices, and political positions. Second, it could result from the ChatGPT training process, where people produce prompt solutions and then sort the resulting responses based on their quality. Third, it could arise from OpenAI's content moderation filters, which are intended to prevent harmful content.

China has taken an active approach to the regulation of artificial intelligence. The regulatory package published in July 2023 also addresses ideology, stating that generative AI services operating in China must adhere to the core values of socialism. AI should not produce content that threatens national sovereignty, security, or unity, damages the nation's image, incites separatism, or spreads terrorism, extremism, or ethnic hatred.

Controlling creative artificial intelligence is challenging, as it can produce surprising content. But when everything happens in digital form, monitoring is also easier and can be automated with the help of artificial intelligence.

Although China wants to regulate and control the use of artificial intelligence, it also wants to ensure that it stays at the forefront of AI development. Therefore, the regulation does not apply to generative AI in cases where it is not offered to end users. Ideology should not hinder innovation.

8.5 Artificial Intelligence in the Hands of a Dictator – A Dangerous Combination

The use of large language models can be both beneficial and potentially dangerous depending on who uses them and for what purpose. We have

[86] Hartmann, Jochen, Schwenzow, Jasper & Witte, Maximilian (2023): The political ideology of conversational AI: Converging evidence on ChatGPT's pro-environmental, left-libertarian orientation. SSRN Electronic Journal. 10.2139/ssrn.4316084. https://arxiv.org/pdf/2301.01768.pdf

discussed in this chapter how artificial intelligence affects our language, values, and worldview. In the wrong hands, artificial intelligence is a dangerous weapon. With it, a dictator could bolster their power in many ways, such as spreading propaganda, manipulating public opinion, enhancing communication surveillance, or smearing their opponents.

Large language models can produce large amounts of text very quickly. This means that a dictator could use the model to produce and spread massive propaganda. The propaganda could be targeted at both their own population and an international audience, with the aim of justifying their own actions and smearing opponents.

Artificial intelligence can be used to create highly targeted communication that is tailored to influence certain types of people or population groups. A dictator could use such technology to shape opinions, steer social discussion in a desired direction, or brainwash people into lies or delusions.

Artificial intelligence enables the analysis of large amounts of data, such as social media messages. A dictator could take advantage of this to infringe on privacy and enhance population surveillance, for example, to identify opponents and critical voices. The dictator could use artificial intelligence to create and spread disinformation that destabilizes opponents or distracts attention from the dictator's own misconduct.

8.6 How do we Preserve our Culture and Values in the AI Revolution?

Artificial intelligence has enormous potential to improve our society at many levels, including supporting democracy. It can help improve the accessibility of information, increase participation and efficiency in decision-making, and support public services.

In the era of artificial intelligence, it is important that we ensure the preservation of our own culture, values, and worldview, taking into account the diversity of society. This requires extensive discussion and cooperation between various parties, such as politicians, other socially significant decision-makers, companies, and developers.

It is also important to invest in technological self-reliance and digital competence, so that we can understand and control the impacts of artificial intelligence. In addition, we must consider our digital sovereignty, i.e., our right to decide on our own digital resources and data protection. Only in this way can we ensure that artificial intelligence serves our values and interests.

Finland has a good starting point for the development of extensive language models, as they have been studied for a long time in many Finnish universities. The first Finnish extensive language model, FinGPT-3, was developed in Turku.[87] Its driving forces, Professor Filip Ginter and researcher Sampo Pyysalo from the University of Turku, are pioneers in language research and language model development in Finland.

In August, Silo AI's subsidiary SiloGen and the Turku University research group TurkuNLP announced they were coordinating a consortium aimed at developing the world's largest open language model. The supercomputer LUMI is being used in the development of the model. The consortium aims to offer this language model in all official European languages.[88]

---

87 Storås, Niclas / HS Visio 19.5.2023: Made in Finland. https://www.hs.fi/visio/art-2000009589999.html

88 Pulliainen, Mikko / Talouselämä 29.8.2023: Researchers from Turku and a Finnish AI company plan to develop the world's largest open language model. https://www.talouselama.fi/uutiset/turkulaistutkijat-ja-suomalainen-tekoaly-yhtio-aikovat-kehittaa-maailman-suurimman-avoimen-kielimallin/1865e8f2-315a-40c2-ba5e-e9e39140ac29

CHAPTER

9

Singularity – AI Surpasses Human Understanding

In early August, I visited the University of Leiden in the Netherlands with my three sons, as the middle one is starting his master's studies there. We sat in the cafe of the oldest botanical garden in the Netherlands with artificial intelligence professor Stefan Leijnen. Around us, we saw magnificent plants – the oldest parts of the garden date back to 1590. Stefan has studied artificial intelligence from many perspectives and also pondered the relationship between artificial intelligence and consciousness.

We end up discussing the singularity, which refers to the level at which artificial intelligence is capable of independent, uncontrollable, and exponential development. Others have also awakened to this discussion, and the topic has generated a number of appeals.

In March 2023, over 30,000 researchers, business leaders, professors, and other individuals concerned about artificial intelligence signed an open letter appealing to all AI laboratories to immediately halt the training of any AI systems more powerful than GPT-4 for at least six months.[89]

A couple of months later, in May, 350 industry leaders signed a brief petition: "Mitigating the risk of extinction caused by artificial intelligence should be a global priority alongside other societal scale risks such as pandemics and nuclear war."[90]

---

[89] Future of Life Institute / An Open Letter 23.3.2023: Pause Giant AI Experiments. https://futureoflife.org/open-letter/pause-giant-ai-experiments/

[90] Center for AI Safety: Statement on AI Risk. https://www.safe.ai/statement-on-ai-risk

Talking about the risks of artificial intelligence is not new, but why are these issues now coming up so strongly in the discussion? The capabilities of ChatGPT, Midjourney, and other AI applications have surprised researchers, business leaders, and the general public. The rapid development of AI is considered a threat. There are various threat scenarios related to AI, some of which may seem like science fiction. One such is the so-called "paperclip problem", where the AI is tasked with making paperclips and it does so from all available material, including humans, thus destroying all of humanity.

The discussion on artificial intelligence is colored by a contradictory paradox. On the one hand, sensible people fear superhuman AI that destroys the world and takes jobs, and on the other hand, they criticize AI's inability to calculate, provide sensible answers, or perform other tasks.

9.1 Weak, Strong, and Super Artificial Intelligence

The "intelligence" of artificial intelligence is often classified as follows: narrow or weak artificial intelligence, general or strong artificial intelligence (AGI, artificial general intelligence) and super intelligence (super intelligence).

Regardless of the development stage of artificial intelligence, it has the following features. Artificial intelligence can decide independently, as it is capable of processing complex data, evaluating different options, and choosing the most suitable course of action based on predefined objectives.

Secondly, artificial intelligence can perform tasks that traditionally require human intelligence, such as language processing, visual perception, or problem-solving. In addition, artificial intelligence is capable of continuously learning and developing. It can acquire new knowledge, develop its models, and improve its performance over time with new data.

How independently artificial intelligence does all this determines the strength and capability of the AI. So, it's a matter of whether it's weak or strong artificial intelligence.

At the moment, artificial intelligence is weak. It is designed to perform a specific task, such as voice recognition, diagnosing device malfunctions, correcting spelling, predicting arrival time in a map application, or choosing which video you want to watch next. Everyday tasks.

Artificial general intelligence (AGI) describes artificial intelligence that has the ability to understand, learn, and apply knowledge to a wide range of tasks at human level or better. In theory, AGI can perform any intellectual task that a human can do. It can make plans, understand complex concepts, learn from experience, adapt to new situations, and demonstrate social intelligence and creativity. Strong artificial intelligence does not yet exist.

But generative AI applications, such as ChatGPT, and the underlying AI models, such as GPT-4, possess strong AI characteristics. For example, the GPT-4 model is capable of performing very different tasks, such as answering questions, writing poems, and diagnosing diseases. This makes it generic in terms of its capabilities.

Since the basic models behind GPT-4 are very complex, we don't even know all the possibilities associated with it yet. In the end, however, it is just weak artificial intelligence, which is guided by a human to perform a certain task. It doesn't perform anything independently and often falls far short of human intelligence.

Strong artificial intelligence may not necessarily be one AI model, but a way in which different AI models are connected and taught to operate independently. Strong artificial intelligence may not emerge in one giant leap, but as a result of several steps.

AutoGPT is an open-source project that aims to build intelligent autonomous agents.[91] The project has many challenges. For example, weak continuous learning, as extensive language models are static, and they do not learn on their own, but they need to be retrained. Taking into account the uncertainty and ambiguity of the real world is difficult, which limits AutoGPT's ability to operate in different environments. Similarly, it has difficulty achieving repetitive self-development based on previous results.

It's easy to improve, say, chess game results by playing it often. But independent continuous self-improvement in many areas where the goals are vague and complex is a challenge of a completely different scale.[92]

To reach human-level thinking, artificial intelligence should also be able to utilize human experience of the physical world. Current language models are based solely on written text, which leads to incorrect information and errors in AI reasoning. Google's researchers have been trying to develop a concept that links language model reasoning to the physical world.[93]

---

[91] GitHub: An experimental open-source attempt to make GPT-4 fully autonomous. https://github.com/Significant-Gravitas/Auto-GPT

[92] Pieri, Lorenzo / Lorenzo Pieri's Blog 2.7.2023: How to Fix AutoGPT and Build a Proto-AGI. https://lorenzopieri.com/autogpt_fix/

[93] Liu, Ruibo, Wei, Jason, Gu, Shixiang Shane, Wu, Te-Yen, Vosoughi, Soroush, Cui, Claire, Zhou, Denny & Dai, Andrew M. (2022): Mind's Eye: Grounded Language Model Reasoning through Simulation. Google Research, Dartmouth College. https://arxiv.org/pdf/2210.05359.pdf

9.2 Superintelligence

According to philosopher Nick Bostrom, superintelligence is any intelligence that surpasses human cognitive performance in practically all areas of interest. OpenAI has also launched the concept of superintelligence, declaring that it will be the most significant technology humanity has ever invented.[94] According to OpenAI, superintelligence is better AI than strong AI, whatever that means. A skeptic could interpret this as an intentional attempt to divert attention to the future, because that's where the danger lies.

Behind superintelligence is the idea that artificial intelligence research can produce software so intelligent that it could program and improve itself – a feature called "recursive self-improvement". It could continue to improve itself in an increasing cycle, leading to superintelligence. This scenario is called an intelligence explosion. Such intelligence would not have human intelligence limitations, and it might be able to invent or discover almost anything.

However, the world is very complex, and even though intelligence seems like a clear concept, it is not. The question is, according to what criteria would artificial intelligence optimize its learning process. Since the goal and its metrics are not clear, artificial intelligence could develop in any direction, like nature or humans.

A highly intelligent person can use their intelligence in many ways. For research and development or even playing video games. Literature is full of theories and scenarios where artificial intelligence ultimately seeks to destroy humanity. Just as well, it could start making art or playing video games. Even if the goal set for artificial intelligence itself might be sensible, it could do harmful things in striving for it.

Even though superintelligence is not in sight, it is still useful to research and develop ways to guide artificial intelligence to make sensible choices, and prevent it from making mistakes in given tasks. Complex systems can be unpredictable and act against our human reasoning. We don't need to fear the dominance of artificial intelligence, as it doesn't have its own will, but its reasoning is based on probabilities. However, we should be concerned about how to get artificial intelligence to function correctly in given tasks, so that it doesn't cause harm.

[94] Leike, Jan & Stuks've, Ilya / OpenAI 5.7.2023: Introducing Superalignment. https://openai.com/blog/introducing-superalignment

Intelligence is not consciousness. Although artificial intelligence is intelligent, it does not possess human-like judgment or consciousness. Consciousness refers to an individual's subjective experience and knowledge of their own existence, their environment, and their own inner mental states. It is the ability to perceive, interpret, and respond to environmental stimuli as well as to reflect on one's own mental states. The ability to think, or even to think about thinking, in the words of philosopher Esa Saarinen.

We do not yet fully understand how human consciousness arises, let alone be able to make artificial intelligence conscious. However, this has not stopped the contemplation and research of artificial intelligence consciousness. The more complex and human-like artificial intelligence systems we develop, the more attention the subject receives. Artificial intelligence may seem conscious, but it is not.

9.3 Singularity: The Turning Point in AI Development

Singularity generally refers to a point in development or event where normal rules and predictions no longer apply. The term originates from astrophysics, specifically cosmology and the study of black holes. A black hole is an extremely densely compressed mass concentration of space-time, and singularity, according to the theory of relativity, is a point formed in the center of a black hole that has no volume at all but consumes all the mass of the black hole. Singularity is a point where the basic laws of nature no longer apply and the properties of space-time become extreme.

Artificial intelligence singularity refers to a hypothetical future point when artificial intelligence systems reach a level of intelligence and capabilities that exceed human possibilities and understanding. Experts' opinions on the likelihood, timeline, and consequences of singularity vary greatly. Some predict the doom of mankind, others a peaceful, mutually reinforcing coexistence of machines and humans.

The concept of singularity was popularized by science fiction writer and mathematician Vernor Vinge in his 1993 article "Technological Singularity".[95] He envisioned a world where artificial intelligence could independently improve and develop itself at an overwhelming pace. After reaching the singularity, the development of artificial intelligence could happen so quickly

[95] Vinge, Vernor / VISION-21 Symposium 30.–31.3.1993: Technological singularity. http://cmm.cenart.gob.mx/delanda/textos/tech_sing.pdf

and its abilities would be so complex that it would be difficult or impossible for humans to predict and understand its operation. Ultimately, this would lead to the destruction of mankind.

Futurist and computer engineer Ray Kurzweil's view of the singularity is more optimistic. He sees that man and machine will merge into a new type of entity, where the capabilities of artificial intelligence enhance human abilities and vice versa. Hybrid artificial intelligence would allow humans to connect directly to the cloud with their brains alone.

The fact that artificial intelligence is approaching singularity does not mean that it will achieve consciousness or be malicious and seek to destroy us. Rather, it's about technology development having progressed so far that we no longer fully understand the complexity of artificial intelligence. This can lead to unpredictable consequences.

When artificial intelligence has spread widely, it can amplify human thoughts very quickly, just as has happened in social media. The impacts of strong artificial intelligence are difficult to predict, and they may be much more harmful than the effects of social media, as artificial intelligence also amplifies evil forces. On the other hand, it is possible that goodness and democracy will prevail.

Predicting all these factors is difficult, and that's why we need to take a responsible approach to the development of artificial intelligence. It's important to create mechanisms that help us understand these complex systems while ensuring that malicious forces do not gain control over artificial intelligence.

That's why we need to strive to develop safe artificial intelligence, whether it's weak or strong – even weak artificial intelligence requires care. It's downright dangerous to focus too much on controlling artificial intelligence in the future, as artificial intelligence is already used in daily decision-making.

9.4 What Could go Wrong?

If the goals of artificial intelligence are not perfectly aligned with human values, it may perform actions that are harmful to humans. This is called the AI alignment problem.

Steering artificial intelligence is a process in which the goals and actions of artificial intelligence are aligned with human values, interests, and goals. By steering artificial intelligence, it can be ensured that artificial intelligence operates in accordance with human interests and promotes societal goals such as justice, safety, and well-being. This requires a multidisciplinary approach that combines perspectives from computer science, ethics, psychology, and other fields.

OpenAI announced in June 2023 that they are developing ways to enable the development of superintelligence in a way that it does not destroy humanity.[96] It sounds dramatic, and it is. In reality, directing artificial intelligence is very practical. Typically, people evaluate the performance of artificial intelligence. As AI models become increasingly capable, they can perform tasks that are difficult or even impossible for humans to evaluate. Therefore, AI is also needed to support human evaluation ability when evaluating the performance of AI.

An example of how artificial intelligence supports human assessment of AI performance is the evaluation of book summaries written by AI. Evaluating the model's results takes a lot of time from humans, as they would first need to read the book and then assess how good the summary was. To facilitate evaluation, OpenAI has trained another AI model that made a summary of smaller parts of the book and a summary of them.[97] This way, a person can read those part summaries and compare them to the whole book summary, which saves time. In this way, it is possible to create a scalable process that ensures the AI model's direction in its task, in this case, creating book summaries.

[96] Altman, Sam / OpenAI 24.2.2023: Planning for AGI and beyond. https://openai.com/blog/planning-for-agi-and-beyond

[97] Wu, Jeffrey, Lowe, Ryan & Leike, Jan / OpenAI 23.9.2021: Summarizing books with human feedback. https://openai.com/research/summarizing-books

Epilogue

This book was created in collaboration with artificial intelligence. AI has helped me refine the book's ideas, explain topics, condense text, and produce images – even the book's cover was created with the help of AI in the hands of Riikka Rajapuro. I also learned a lot about the concrete use of AI and when it doesn't work.

At the end of the book, it's time to crystallize and recap the key ideas that we have gone through on our journey to the heart of the AI revolution.

The book kicks off with a deep dive into the world of generative artificial intelligence. Together, we have explored extensive artificial intelligence models, studied their operating principles and instructions through text, images, and videos. We have examined the impact of artificial intelligence on working life, its opportunities for increasing productivity, and its effects on society as a whole. We have also delved into the development of robotics, examined the legal questions raised by artificial intelligence, and pondered the impact of artificial intelligence on language, culture, and worldview. Finally, we have envisioned the future and the kind of world we might be heading towards.

The AI revolution is not just a technological change, but it radically renews our entire society from work to productivity and from law to culture. It is a new civic skill, the mastery of which is essential for individuals, businesses, and society alike. As we have seen, AI can create new opportunities, but it also brings with it risks that we must learn to manage.

I would like to challenge every reader to examine their own relationship with artificial intelligence: How is it already a part of your life and how could you utilize it even better? Is your company ready to take advantage of the opportunities brought by artificial intelligence? How can you use it to affect productivity? It is also important to consider how we can together preserve and protect our values in this change.

Even though we have already achieved a lot, we are only at the beginning of the development of artificial intelligence. We must continue forward, as even greater changes await us in the future.

I have been involved in the development of artificial intelligence from the very beginning, and following the development has been really exciting. It has been interesting to see how artificial intelligence has changed from an abstract concept to a concrete tool that shapes our lives in many ways.

Finally, I want to thank you, the reader, for this journey into the heart of the AI revolution. I hope this book has provided you with new insights, thoughts, and tools to bravely face the future. Thank you for being with me on this journey. And remember, the AI revolution is just the beginning. Keep learning, stay curious, and actively shape the future – the future is in our hands.

Sources

Acar, Oguz A. / Harvard Business Review 6.6.2023: AI prompt engineering isn't the future. https://hbr.org/2023/06/ai-prompt-engineering-isnt-the-future

Altman, Sam / OpenAI 24.2.2023: Planning for AGI and beyond. https://openai.com/blog/planning-for-agi-and-beyond

Armstrong, Evan / Every 18.5.2023: Meet Your New Finance Intern: GPT-4. How AI will reshape what finance professionals do. https://every.to/napkin-math/meet-your-new-finance-intern-gpt-4

Armstrong, Evan / Napkin Math 22.6.2023: AI Can Construct a Trading Strategy Now. https://every.to/napkin-math/ai-can-construct-a-trading-strategy-now

Barker, Nat / Deezen 26.4.2023: ZHA developing "most" projects using AI-generated images says Patrik Schumacher. https://www.dezeen.com/2023/04/26/zaha-hadid-architects-patrik-schumacher-ai-dalle-midjourney/

About Belatchew Architectural Firm's AI project on the website of Vinnova, the Swedish government's technology funding organization: https://www.vinnova.se/en/p/creaitor---creative-ai-for-innovative-and-sustainable-architecture/

Bloomberg 30.3.2023: Introducing BloombergGPT, Bloomberg's 50-billion parameter large language model, purpose-built from scratch for finance. https://www.bloomberg.com/company/press/bloomberggpt-50-billion-parameter-llm-tuned-finance/

Bradford, Anu / Foreign Affairs 27.6.2023: The Race to Regulate Artificial Intelligence. https://www.foreignaffairs.com/united-states/race-regulate-artificial-intelligence?check_logged_in=1&utm_medium=promo_email&utm_source=lo_flows&utm_campaign=registered_user_welcome&utm_term=email_1&utm_content=20230708

Briggs, Joseph & Kodnani, Devesh / Goldman Sachs 26.3.2023: The Potentially Large Effects of Artificial Intelligence on Economic Growth. Global Economics Analyst. https://www.gspublishing.com/content/research/en/reports/2023/03/27/d64e052b-0f6e-45d7-967b-d7be35fabd16.html

Brittain, Blake / Reuters 23.2.2023: AI-created images lose U.S. copyrights in test for new technology. https://www.reuters.com/legal/ai-created-images-lose-us-copyrights-test-new-technology-2023-02-22/

Brynjolfsson, Erik, Li, Danielle & Raymond, Lindsey R. (2023): Generative AI at Work. NBER. https://www.nber.org/papers/w31161

Cao, Yong, Zhou, Li, Lee, Seolhwa, Cabello, Laura, Chen, Min & Hershcovich, Daniel (2023): Assessing Cross-Cultural Alignment between ChatGPT and Human Societies: An Empirical Study. https://arxiv.org/abs/2303.17466

Center for AI Safety: Statement on AI Risk. https://www.safe.ai/statement-on-ai-risk

ChatGPT – Release Notes 9.8.2023: https://help.openai.com/en/articles/6825453-chatgpt-release-notes

Cho, Winston / The Hollywood Reporter 29.6.2023: Authors Sue OpenAI Claiming Mass Copyright Infringement of Hundreds of Thousands of Novels. https://www.hollywoodreporter.com/business/business-news/authors-sue-OpenAI -novels-1235526462/

Chui, Michael, Hazan, Eric, Roberts, Roger, Singla, Alex, Smaje, Kate, Sukharevsky, Alex, Yee, Lareina & Zemmel, Rodney / McKinsey Digital 14.5.2023: The economic potential of generative AI: The next productivity frontier. https://www.mckinsey.com/capabilities/mckinsey-digital/our-insights/the-economic-potential-of-generative-ai-the-next-productivity-frontier#introduction

Coldewey, Devin / TechCrunch 29.12.2022: Meta's Make-A-Video AI achieves a new, nightmarish state of the art. https://techcrunch.com/2022/09/29/meta-make-a-video-ai-achieves-a-new-creepy-state-of-the-art/

Eapen, Tojin T. , Finkenstadt, Daniel J., Folk, Josh & Venkataswamy, Lokesh / Harvard Business Review July–August 2023: How Generative AI Can Augment Human Creativity. https://hbr.org/2023/07/how-generative-ai-can-augment-human-creativity

European Commission 21.4.2021: Proposal for a Regulation of the European Parliament and the Council on harmonised rules on artificial intelligence (AI regulation) and amending certain Union regulations. https://data.consilium.europa.eu/doc/document/ST-8115-2021-INIT/fi/pdf

European Commission: Proposal on regulatory framework for artificial intelligence | Shaping Europe's digital future, updated 20.6.2023. https://digital-strategy.ec.europa.eu/fi/policies/regulatory-framework-ai

European Parliament News 8.6.2023: EU AI Act: first regulation on artificial intelligence. https://www.europarl.europa.eu/news/en/headlines/society/20230601STO93804/eu-ai-act-first-regulation-on-artificial-intelligence

European Parliament: Amendments adopted by the European Parliament on 14 June 2023 on the proposal for a regulation of the European Parliament and of the Council on laying down harmonised rules on artificial intelligence

(Artificial Intelligence Act) and amending certain Union legislative acts (COM(2021)0206 – C9-0146/2021 – 2021/0106(COD)). https://www.europarl.europa.eu/doceo/document/TA-9-2023-0236_EN.html

Future of Life Institute / An Open Letter 23.3.2023: Pause Giant AI Experiments. https://futureoflife.org/open-letter/pause-giant-ai-experiments/

Getty Images Statement 17.1.2023. https://newsroom.gettyimages.com/en/getty-images/getty-images-statement

GitHub: An experimental open-source attempt to make GPT-4 fully autonomous. https://github.com/Significant-Gravitas/Auto-GPT

Grieshaber, Kirsten / Associated Press (AP) June 10, 2023: Can a chatbot preach a good sermon? Hundreds attend church service generated by ChatGPT to find out. https://apnews.com/article/germany-church-protestants-chatgpt-ai-sermon-651f21c24cfb47e3122e987a7263d348

Hartmann, Jochen, Schwenzow, Jasper & Witte, Maximilian (2023): The political ideology of conversational AI: Converging evidence on ChatGPT's pro-environmental, left-libertarian orientation. SSRN Electronic Journal. 10.2139/ssrn.4316084. https://arxiv.org/pdf/2301.01768.pdf.

Helsingin Sanomat 23.12.2022. Reader's opinion: It's clear that artificial intelligence changes the nature of knowledge work: https://www.hs.fi/mielipide/art-2000009280423.html.

John / AI for Folks 11.7.2023: How to Give ChatGPT a Table for Basic Data Processing and Analysis. https://aiforfolks.com/how-to-give-chatgpt-a-table-for-basic-data-processing-and-analysis/

University of Jyväskylä / For the student / Rules and regulations guiding studies: Use of AI-based applications in studies 8.6.2023. https://www.jyu.fi/fi/opiskelijalle/opintoja-ohjaavat-saadokset/tekoalypohjaisten-sovellusten-kaytto-opiskelussa-jyu-ohjeet-ja-linjaukset

Kalliamvakou, Eirini / GitHub 7.9.2022: Research: quantifying GitHub Copilot's impact on developer productivity and happiness. https://github.blog/2022-09-07-research-quantifying-github-copilots-impact-on-developer-productivity-and-happiness/

Kanerva, Jenna, Kilpeläinen, Jemina, Kupari, Hanna-Mari, Saarni, Jenna & Skantsi, Valtteri / University of Turku blog 31.8.2021: Towards linguistically diverse artificial intelligence. https://blogit.utu.fi/utu/2021/08/31/kohti-kielellisesti-monimuotoista-tekoalya/

Kauppalehti 9.3.2023. Opinion piece: The brand is an algorithm - the use of artificial intelligence models changes the way a brand is implemented. https://www.kauppalehti.fi/uutiset/brandi-on-algoritmi-tekoalymallien-kaytto-muuttaa-tapaa-jolla-brandi-toteutetaan/47a26e18-b850-4da4-8bed-22f6e0c19f83

Kemper, Jonathan / The Decoder 19.6.2023: ChatGPT Guide: Use these prompt strategies to maximize your results. https://the-decoder.com/chatgpt-guide-prompt-strategies/

Knight, Will / Wired 19.8.2022: Algorithms Can Now Mimic Any Artist. Some Artists Hate It. https://www.wired.com/story/artists-rage-against-machines-that-mimic-their-work/

Kremp, Moritz in communication service X on May 30, 2023: https://twitter.com/moritzkremb/status/1663531203071442944?s=61&t=Zk98bDqXMraO3VYJypazWA

Landi, Heather / Fierce Healthcare 25.5.2023: Epic is going all in on generative AI in healthcare. Here's why health systems are eager to test-drive it. https://www.fiercehealthcare.com/health-tech/epic-moves-forward-bring-generative-ai-healthcare-heres-why-handful-health-systems-are

Leike, Jan & Sutskever, Ilya / OpenAI 5.7.2023: Introducing Superalignment. https://openai.com/blog/introducing-superalignment

Liu, Ruibo, Wei, Jason, Gu, Shixiang Shane, Wu, Te-Yen, Vosoughi, Soroush, Cui, Claire, Zhou, Denny & Dai, Andrew M. (2022): Mind's Eye: Grounded Language Model Reasoning through Simulation. Google Research, Dartmouth College. https://arxiv.org/pdf/2210.05359.pdf

Machine Learning Mastery 16.6.2023: Prompt Engineering for Effective Interaction with ChatGPT.https://machinelearningmastery.com/prompt-engineering-for-effective-interaction-with-chatgpt/

Madiega, Tambiama / European Parliamentary Research Service 6/2023: Artificial intelligence act. https://www.europarl.europa.eu/RegData/etudes/BRIE/2021/698792/EPRS_BRI(2021)698792_EN.pdf

McKinsey Digital 14.6.2023: The economic potential of generative AI: The next productivity frontier. https://www.mckinsey.com/capabilities/mckinsey-digital/our-insights/the-economic-potential-of-generative-AI-the-next-productivity-frontier#/

Milmo, Dan & agency / The Guardian 23.6.2023: Two US lawyers fined for submitting fake court citations from ChatGPT. https://www.theguardian.com/technology/2023/jun/23/two-us-lawyers-fined-submitting-fake-court-citations-chatgpt

Mok, Aaron / Insider 1.3.2023: 'Prompt engineering' is one of the hottest jobs in generative AI. Here's how it works. https://www.businessinsider.com/prompt-engineering-ai-chatgpt-jobs-explained-2023-3?r=US&IR=T.

Murgia, Madhumita / Financial Times 14.4.2023: OpenAI 's red team: the experts hired to 'break' ChatGPT. https://www.ft.com/content/0876687a-f8b7-4b39-b513-5fee942831e8

Nicolaou, Anna / Financial Times 11.4.2023: Streaming services urged to clamp down on AI-generated music. https://www.ft.com/content/aec1679b-5a34-4dad-9fc9-f4d8cdd124b9

Ojanperä, Tero & Mokko, Kari / HS guest column 7.12.2016: "Finland needs an artificial intelligence strategy", https://www.hs.fi/paivanlehti/07122016/art-2000004897243.html

OpenAI 14.3.2023: GPT-4. https://OpenAI .com/research/gpt-4

OP Group 2022: Economic Balance - Artificial Intelligence Transparency Report. https://www.op.fi/documents/20556/39794367/OP-Ryhm%C3%A4-Tekoalyn-lapinakyvyysraportti-Talouden-tasapaino-2022.pdf/8403ea51-e911-b40d-a1dd-33d25aff2d93

OP Group / Cision 21.1.2023: OP implements a human-centered AI assessment model and publishes its first transparency report on its AI. https://news.cision.com/fi/op-ryhma/r/op-ottaa-kayttoon-ihmislahtoisen-tekoalyn-arviointimallin-ja-julkaisee-ensimmaisen-lapinakyvyysrapor,c3696412

Parti, Ayush / Product Hunt/Stories 6.6.2023: Midjourney vs. Stable Diffusion vs. DALL-E: What's best in 2023? https://www.producthunt.com/stories/midjourney-vs-stable-diffusion-vs-dall-e

Peng, Sida, Kalliamvakou, Eirini, Cihon, Peter & Demirer, Mert 13.2.2023: The Impact of AI on Developer Productivity: Evidence from GitHub Copilot. 2302.06590.pdf (arxiv.org)

Pieri, Lorenzo / Lorenzo Pieri's Blog 2.7.2023 : How to Fix AutoGPT and Build a Proto-AGI. https://lorenzopieri.com/autogpt_fix/

Prompt Warrior website: https://www.thepromptwarrior.com/

Pulliainen, Mikko / Talouselämä 29.8.2023: Researchers from Turku and a Finnish AI company plan to develop the world's largest open language model. https://www.talouselama.fi/uutiset/turkulaistutkijat-ja-suomalainen-tekoaly-yhtio-aikovat-kehittaa-maailman-suurimman-avoimen-kielimallin/1865e8f2-315a-40c2-ba5e-e9e39140ac29

PwC March 15, 2023: PwC announces strategic alliance with Harvey, positioning PwC's Legal Business Solutions at the forefront of legal generative AI. https://www.pwc.com/gx/en/news-room/press-releases/2023/pwc-announces-strategic-alliance-with-harvey-positioning-pwcs-legal-business-solutions-at-the-forefront-of-legal-generative-ai.html

Raieli, Salvatore / Medium 17.3.2023: Google Med-PaLM: The AI Clinician. https://towardsdatascience.com/google-med-palm-the-ai-clinician-a4482143d60e

Ramezani, Aida & Xu, Yang (2023): Knowledge of cultural moral norms in large language models. arXiv preprint arXiv:2306.01857: https://aclanthology.org/2023.acl-long.26.pdf

Reed, Rachel / Harvard Law Today 2.5.2023: AI created a song mimicking the work of Drake and The Weeknd. What does that mean for copyright law? https://hls.harvard.edu/today/ai-created-a-song-mimicking-the-work-of-drake-and-the-weeknd-what-does-that-mean-for-copyright-law/

Reinholdt, Eric / 30X40 Design Workshop 30.5.2023: Using AI as design tool in my architecture practice. https://thirtybyforty.com/blog/using-ai-as-design-tool-in-my-architecture-practice

Repo, Liina, Skantsi, Valtteri, Rönnqvist, Samuel, Hellström, Saara, Oinonen, Miika, Salmela, Anna, Biber, Douglas, Egbert, Jesse, Pyysalo, Sampo &

Laippala, Veronika (2021): Beyond the English Web: Zero-Shot Cross-Lingual and Lightweight Monolingual Classification of Registers. In Proceedings of the 16th Conference of the European Chapter of the Association for Computational Linguistics: Student Research Workshop, pages 183–191, Online. Association for Computational Linguistics. https://aclanthology.org/2021.eacl-srw.24.pdf

Savkin, Aleksey / Medium 27.12.2022: ChatGPT AI for Scenario Planning: Generate Plausible Stories. https://medium.com/@bscdesigner/chatgpt-ai-for-scenario-planning-generate-plausible-stories-34b7730da070

Schuberg, Charlotte / Geekwire 30.3.2023: Microsoft Research head Peter Lee on the applications of GPT-4 in medicine and life sciences. https://www.geekwire.com/2023/microsoft-research-head-peter-lee-on-the-implications-of-gpt-4-for-medicine-and-research/

Siemon, Dominik & Kedziora, Damian / Hawaii International Conference on System Sciences (HICSS-56), January 2023: From Accountant to Software Developer - Transforming Employees with Robotic Process Automation (RPA). https://www.researchgate.net/publication/364139508_From_Accountant_to_Software_Developer_-_Transforming_Employees_with_Robotic_Process_Automation_RPA

Spicer, Tracey / The Guardian 23.2.2023: AI art replicates inequity at scale. we need to learn about its biases – and outsmart the algorithm. https://www.theguardian.com/books/2023/feb/24/ai-art-replicates-inequity-at-scale-we-need-to-learn-about-its-biases-and-outsmart-the-algorithm

Stacey, Stephanie / TechMonitor 17.4.2023: Deepfakes are being democratised — and getting harder to detect. https://techmonitor.ai/technology/ai-and-automation/deepfakes-are-being-democratised-and-getting-harder-to-detect

Steiny, Sam & Trisal, Shiv / Databricks 30.3.2023: The Great Unlock: Large Language Models in Manufacturing. https://www.databricks.com/blog/great-unlock-large-language-models-manufacturing

Storytel 13.6.2023: Storytel enters strategic partnership with ElevenLabs and announces upcoming launch of new VoiceSwitcher feature. https://investors.storytel.com/en/storytel-enters-strategic-partnership-with-elevenlabs-and-announces-upcoming-launch-of-new-voiceswitcher-feature/

Storås, Niclas / HS Visio 19.5.2023: Made in Finland. https://www.hs.fi/visio/art-2000009589999.html

St. Pierre, Nick communication service X on July 25, 2023: https://twitter.com/nickfloats/status/1683664356356005889

Teosto (2021): Teosto takes care of Sibelius's works' copyrights: https://www.teosto.fi/teosto-huolehtii-sibeliuksen-teosten-tekijanoikeuksista/

The Cleverest 31.8.2022: Using DALL-E 2 for interior design: https://thecleverest.com/using-dall-e-2-for-interior-design/

Tietoevry 29.6.2023: Tietoevry accelerates Generative AI innovation – Tietoevry Care and Silo AI are developing a Finnish speaking AI assistant for healthcare professionals. https://www.tietoevry.com/en/newsroom/all-news-and-releases/press-releases/2023/06/tietoevry-accelerates-generative-ai-innovation-tietoevry-care-and-silo-ai-are-developing-a-finnish-speaking-ai-assistan/

Touvron, H., Lavril, T., Izacard, G., Martinet, X., Lachaux, M.-A., Lacroix, T., Rozière, B., Goyal, N., Hambro, E., Azhar, F., Rodriguez, A., Joulin, A., Grave, E. & Lample, G. / Cornell University 27.2.2023: LLaMA: Open and Efficient Foundation Language Models. https://arxiv.org/abs/2302.13971

Tuomi, Ilkka / HS Opinion 3.6.2023: Artificial intelligence cannot be left in the school's mobile phone park. https://www.hs.fi/mielipide/art-2000009622272.html.

United States Court of Appeals 10.2.2000: Sony Computer Entertainment v. Connectix Corp., 203 F.3d 596 (9th Cir. 2000) https://casetext.com/case/sony-computer-entertainment-v-connectix-corp-2

Verma, Pranshu & De Vynck, Gerrit / The Washington Post 2.6.2023: ChatGPT took their jobs. Now they walk dogs and fix air conditioners. https://www.washingtonpost.com/technology/2023/06/02/ai-taking-jobs/

Vincent, James / The Verge 15.11.2022: The scary truth about AI copyright is nobody knows what will happen next. https://www.theverge.com/23444685/generative-ai-copyright-infringement-legal-fair-use-training-data

Vinge, Vernor / VISION-21 Symposium 30.–31.3.1993: Technological singularity. http://cmm.cenart.gob.mx/delanda/textos/tech_sing.pdf

Volpicelli, Gian / Politico 23.3.2023: ChatGPT broke the EU plan to regulate AI. https://www.politico.eu/article/eu-plan-regulate-chatgpt-openai-artificial-intelligence-act/

Wiggers, Kyle / TechCrunch 22.12.2022: A brief history of diffusion, the tech at the heart of modern image-generating AI. https://techcrunch.com/2022/12/22/a-brief-history-of-diffusion-the-tech-at-the-heart-of-modern-image-generating-ai/?guccounter=1

Wu, Jeffrey, Lowe, Ryan & Leike, Jan / OpenAI 23.9.2021: Summarizing books with human feedback. https://openai.com/research/summarizing-books

Yoon, Songyee / Games Beat 25.4.2023: Artificial generative intelligence risks a return to cultural colonialism. https://venturebeat.com/games/artificial-generative-intelligence-risks-a-return-to-cultural-colonialism/

Index

For Product Safety Concerns and Information please contact our EU representative GPSR@taylorandfrancis.com Taylor & Francis Verlag GmbH, Kaufingerstraße 24, 80331 München, Germany